T0330624

Higher Education for Sustainable Development Goals

RIVER PUBLISHERS SERIES IN MANAGEMENT SCIENCES AND ENGINEERING

The River Publishers Series in Management Sciences and Engineering aims at publishing high quality books on management sciences and engineering. The Series provides discussion and facilitates exchange of information on principles, strategies, models, techniques, methodologies and applications of management sciences and engineering in the field of industry, commerce, government and services. The Series aims to communicate the latest developments and thinking on the management and engineering subject world-wide. Thus, the main aim of this book series is to serve as a useful reference and to provide a channel of communication to disseminate knowledge between academics, researchers, managers and engineers and other professionals.

The Series seeks to link management sciences and engineering disciplines to helping organizations in private and public sectors addressing technological, business and societal challenges in a context of complexity, uncertainty and change. The Series stimulates the development of new approaches to management and engineering in a human, organizational and societal context, from a socio-technical systems perspective, and in aspects as design, innovation, planning, coordination, communication, engagement and decision-making. Examples of societal challenges requiring new approaches to management and engineering include the sustainable, secure and responsible development of energy, transport, water, infrastructural and informational resources, infrastructures and services. The Series is highlighting cultural and geographical diversity in studies oriented to transformation in organizations and work environments. It emphasizes the role of human resource management, the role of organizational decision-making and collaboration, empowerment of stakeholders, and the management of innovation and engineering activities in a context of organizational change. This way it reflects the diversity of human, organizational and societal conditions.

Books published in the series include research monographs, edited volumes, handbooks and text books. The books provide professionals, researchers, educators, and advanced students in the field with an invaluable insight into the latest research and developments.

Topics covered in the series include, but are by no means restricted to the following:

- Human Resources Management
- Socio-technical Systems Change
- Large-scale Systems in a Human and Societal Context
- Culture and Organizational Behaviour
- Higher Education for Sustainability
- Management in SMEs, large organizations, networks
- Strategic Management
- Entrepreneurship and Business Strategy
- Interdisciplinary Management
- Management and Engineering Education
- Knowledge Management
- Management of Innovation and Engineering
- Operations Strategy, Planning and Decision Making
- Sustainable Management and Engineering
- Production and Industrial Engineering
- Materials and Manufacturing Processes
- Manufacturing Engineering
- Interdisciplinary Engineering
- Management of System Transitions (energy, water, transport)

For a list of other books in this series, visit www.riverpublishers.com

Higher Education for Sustainable Development Goals

Editors

Carolina Machado
University of Minho, Portugal

João Paulo Davim
University of Aveiro, Portugal

Published 2022 by River Publishers
River Publishers
Alsbjergvej 10, 9260 Gistrup, Denmark
www.riverpublishers.com

Distributed exclusively by Routledge
4 Park Square, Milton Park, Abingdon, Oxon OX14 4RN
605 Third Avenue, New York, NY 10017, USA

Higher Education for Sustainable Development Goals / by Carolina Machado, João Paulo Davim.

Routledge is an imprint of the Taylor & Francis Group, an informa business

ISBN 978-87-7022-431-4 (print)
ISBN 978-10-0077-426-9 (online)
ISBN 978-1-003-33303-6 (ebook master)

Contents

Preface xi

List of Figures xiii

List of Tables xvii

List of Contributors xix

List of Abbreviations xxiii

1 Pedagogic Resonance and Threshold Concepts to Access the Hidden Complexity of Education for Sustainability 1
Paulo R. M. Correia, and Ian M. Kinchin
1.1 Mind the Gap Between Pedagogy and Teaching Methods . . 2
1.2 Learning as a Climbing Adventure 5
1.2.1 Disjuncture and The Need for Pedagogic Resonance 6
1.2.2 Climbing the Knowledge Mountains Using Segmented and Hierarchical Learning 8
1.3 Mapping the Conceptual Terrain 10
1.3.1 Threshold Concepts and Changes in Understanding 10
1.3.2 Knowledge Integrations at The Threshold Concepts 13
1.4 Charting the EfS Knowledge Mountains 15
1.5 Implications for Mountain Guides and Adventurers 18
1.6 Acknowlegdements 19
References 19

2 Universities – Players in the Race for Sustainable Development **23**
M. Panait, E. Hysa, M. G. Petrescu, and H. Fu
2.1 Introduction 24
2.2 The Academic Revolutions and Multiple Functions of Universities 26
2.3 Moving the SDG 2030 Agenda Forward Through Higher Education 28
2.4 Responsible Management Education and Sustainable Development 31
2.4.1 Sustainable Development Through Curricula: Some Examples 33
2.4.2 Sustainable Development Through Green Campus Activities: Some Examples 34
2.4.3 Sustainable Development Through Other Initiatives of Universities 35
2.5 Conclusion 36
References 37

3 Sustainability in Portuguese Higher Education Institutions **43**
Ana Caria, Cristiana Leal, Carolina Machado, Benilde Oliveira, and Lídia Oliveira
3.1 Introduction 44
3.2 Higher Education Institutions and Sustainability 45
3.3 Sustainability and HEIs in Portugal 50
3.4 Portuguese HEIs in Sustainability Rankings 55
3.4.1 Sustainability Tracking, Assessment and Rating System 55
3.4.2 Times Higher Education Impact Rankings 56
3.4.3 GreenMetric World Universities Rankings 62
3.5 The Sustainability Report: the Cases of Four Portuguese HEIs 62
3.5.1 Faculty of Engineering of the University of Porto . . 64
3.5.2 University of Minho 66
3.5.3 University of Coimbra 68
3.5.4 ISCTE-University Institute of Lisbon 70
3.6 Conclusion 73
References 74

4 Where is the Brazilian Higher Education Within the Sustainable Development Goal 4? 83
Sidney L. M. Mello, Carlos E. Bielschowsky, Marcelo J. Meriño, and Thaís N. da R. Sampaio
4.1 Introduction . . . 84
4.2 Expansion of HE in Brazil . . . 87
4.2.1 Social Inclusion and Access Equity . . . 91
4.2.2 Professional Training in HE . . . 94
4.2.3 Quality Assessment . . . 97
4.3 The Boom of Distance Learning . . . 100
4.4 Looking into The Future . . . 104
References . . . 109

5 Incorporating SDG 11 in Higher Education Teaching – The Relevance of Mobility on Sustainable Cities and Communities 117
Margarida C. Coelho
5.1 Introduction . . . 118
5.2 Incorporating Mobility in the Teaching and Research Activities of an HEI . . . 120
5.2.1 The Course 'Energy, Mobility and Transportation' . 121
5.2.2 Micromodules . . . 133
5.2.3 Incorporate Students in Research Activities . . . 134
5.3 The Power of Internationalization Under the Teaching Process . . . 136
5.4 Conclusion . . . 138
5.5 Acknowlegdements . . . 139
References . . . 139

6 Push and Pull: Sustainability Education for 21st Century Engineers 141
Salma Shaik, Lakshika N. Kuruppuarachchi, and Matthew J. Franchetti
6.1 Introduction . . . 142
6.1.1 History of Sustainable Development . . . 142
6.1.2 Education for Sustainable Development (ESD) . . . 144
6.2 Push from Students . . . 146
6.2.1 Youth Sustainability Activism . . . 146
6.2.2 Student Sustainability Initiatives . . . 146

6.3 Pull from Industry 147
6.3.1 Involvement of Environmental Engineering in Industries 148
6.3.2 Environmental Engineer vs Sustainability Engineer 148
6.3.3 Role of an Environmental Engineer 148
6.3.4 Distribution of Environmental Engineers in Different Industries 148
6.3.5 Industry Sustainability Practices – Case Studies . . . 149
6.3.6 Environmental Engineering: A New Major 150
6.3.7 Entrepreneurial Thinking Through 3C's 151
6.4 21st Century Engineer 152
6.4.1 Challenges of 21st Century 152
6.4.2 Requirements of a 21st Century Engineer 153
6.5 Sustainability Education in the 21st Century 154
6.5.1 Role of Engineers in Sustainable Development . . . 156
6.5.2 Engineering Education for Sustainable Development (EESD) 156
6.6 Conclusion 158
References 159

7 Unleashing Emotions: The Role of Emotional Intelligence Among Students in Upholding Sustainable Development Goals 171
Christabel Odame, Mrinalini Pandey, and David Boohene
7.1 Introduction 172
7.2 Review of Literature 173
7.2.1 Higher Education and Sustainable Development Goals 173
7.2.2 Emotional Intelligence and Sustainable Development Goals 173
7.2.3 Emotional Intelligence and Environmental Issues . . 174
7.2.4 Emotional Intelligence and Society or Social Issues . 174
7.2.5 Emotional Intelligence and Economy or Financial Issues 175
7.2.6 Methodology 175
7.3 Results 175
7.4 Discussion 176
7.5 Conclusion and Recommendation for Future Research . . . 177
References 178

8 Pedagogy for Living in Harmony with Nature – Sustainability in Higher Education 181
Qudsia Kalsoom, and Sibte Hasan
8.1 Introduction 182
8.2 Capitalism and Human Behaviour 185
8.2.1 Capitalism and Communication Industries 186
8.2.2 Plunder of Nature 188
8.2.3 Pedagogy of Domination/ Consumerism 189
8.2.4 Social Unsustainability 192
8.3 Higher Education for Sustainable Development Goals (SDGs) 193
8.3.1 Pedagogy of Harmony with Nature (PHN) 194
8.4 Conclusion 196
References 197

Index 203

About the Editors 211

Preface

This book cover the issues related to advances in higher education for sustainable development goals. Nowadays, sustainable development is an important concept in higher education. One of the most widely recognized definition is based on *Brundtland* report as *"development that meets the needs of the present without compromising the ability of future generations to meet their own needs."* The three core pillars of sustainable development are environment, society, and economy. Currently, higher education in the context of sustainable development goals (SDGs) is a great challenge. The information about higher education for sustainable development presents a great interest to improve communication between, professors, researchers, and students in universities, institutes, colleges, etc. At this moment, the scientific interest in higher education for sustainable development goals is evident for many important universities in the world.

This research book focuses on all aspects of higher education for sustainable development goals, namely, no poverty, zero hunger, good health and wellbeing, quality education, gender equality, clean water and sanitation, affordable and clean energy, decent work and economic growth, industry, innovation, and infrastructure, reduced inequalities, sustainable cities and communities, responsible consumption and production, climate action, life bellow water, life on land, peace, justice and strong institutions, partnerships.

Higher Education for Sustainable Development Goals contributes to and increases the discussion and debate on the role of Higher Education Institutions in the creation and dissemination of knowledge about sustainability and sustainable development goals, since they are key players in the construction and development of professionals of today, and particularly, of tomorrow.

The role of these professionals, namely with regard to the assimilation and internalization of the importance of the SDGs for today and tomorrow's society is of great relevance. Indeed, teachers, students, among other professionals, have the opportunity to obtain academic education at a higher level in order to practice and develop their professions and support societies.

Also HEIs undertake fundamental and applied research in the different fields improving our understanding of life. They are seen as drivers for the achievement of the full set of goals, through their critical role in human training, knowledge production as well as innovation.

Following these concerns, this book looks to cover the field of *Higher Education for Sustainable Development Goals* in eight chapters. Chapter 1 focuses on *"Pedagogic resonance and threshold concepts to access the hidden complexity of Education for Sustainability,"* Chapter 2 speaks about *"Universities – players in the race for sustainable development."* Chapter 3 covers *"Sustainability in Portuguese Higher Education Institutions,"* while Chapter 4 contains information about "*Where is the Brazilian higher education within the Sustainable Development Goal 4?*" Chapter 5 deals with *"Incorporating SDG 11 in Higher Education Teaching – The relevance of mobility on sustainable cities and communities"*. Chapter 6 highlights *"Push and Pull: Sustainability Education for 21st Century Engineers."* Finally, Chapter 7 covers "*Unleashing Emotions: The Role of Emotional Intelligence among students in upholding Sustainable Development Goals*", and Chapter 8 focuses "*Pedagogy for Living in Harmony with Nature – Sustainability in Higher Education.*"

Aiming to provide some recent research advances on higher education for SDGs field, *Higher Education for Sustainable Development Goals* can be used in final undergraduate courses (namely, engineering, management, among others) or as a subject on higher education for sustainability at the postgraduate level. Also, this book can serve as a useful reference for academics, researchers, engineers, managers as well as all type of professionals related with higher education for sustainable development.

The Editors acknowledge their gratitude to River Publishing for this opportunity and for their professional support. Finally, we would like to thank to all chapter authors for their interest and availability to work on this project.

Carolina Machado
Braga, Portugal

J. Paulo Davim
Aveiro, Portugal

List of Figures

Figure 1.1 Mapping the hidden complexity of the educational approach: (a) iceberg metaphor to highlight that only the tip is visible to external observation and (b) the sophisticated instructional discourse framed from Novak's basic elements of any educational event grey boxes. 4

Figure 1.2 Learning as changes in knowledge structures: (a) mapping the differences among spokes, chains and nets and (b) the disjuncture as a daunting event to avoid rote learning. 7

Figure 1.3 Develop segmented and hierarchical knowledge structures is a strategy to conquer the knowledge mountains. The arrows 1–4 indicate base camps that serve as preparation (conceptual stasis) before climbing the highest peaks (conceptual change). . . 9

Figure 1.4 Overcoming threshold concepts flatten the landscape, and experts struggle to understand the learners' difficulties with the apparent flat terrain (already known). The base camps for the next climbing (arrows 3–4) are visible because they help to enter the 'yet-to-be known'. 12

Figure 1.5 The integrative role of threshold concepts (grey boxes) to overcome the peaks in the knowledge mountain range. Four different integrations are shown considering procedural knowledge (I), upper-tree hierarchy (conceptual knowledge from scientific disciplines, II) and network hierarchy (conceptual knowledge from humanities, III). 13

Figure 1.6	A preliminary nested hierarchy of threshold concepts (white boxes) to access the hidden complexity of EfS. Grey boxes detail the relevance of 'virtue ethics', 'wicked problems' and 'academic literacy' to understand 'sustainability'.	16
Figure 2.1	Mission of universities.	26
Figure 2.2	Steps for university internationalisation.	28
Figure 2.3	Integrated Sustainable Development Goals.	29
Box 2.1	Principles for Responsible Management Education.	32
Figure 3.1	The DESD Milestones.	46
Graph 3.1	Average Score of Portuguese HEIs by SDG and year, in the THE Impact Rankings.	61
Figure 4.1	Growth of higher education between 1999 and 2019, displaying public and private sectors and on-campus and distance learning.	89
Figure 4.2	Percentual growth enrollments in distance learning between 1999 and 2019. The dark line shows the total evolvement, and the light line the private HE progression of enrollments.	90
Figure 4.3	Percentage of graduates between 2009 and 2011 and between 2017 and 2019 by family income profile.	92
Figure 4.4	Total number of high education institutions by category.	96
Figure 4.5	Total number of students enrolled in distance learning courses between the years 1999 and 2019.	101
Figure 4.6	Percentage of students per age (18–60 years old) who graduated on-campus and DL between 2017 and 2019.	102
Figure 4.7	Enade of graduating students of pedagogy course in two HEIs, presential and DL modalities. Each learning modality represents 25% of the total, and Enade grades vary from 5 to 95.	104
Figure 5.1	Demonstration of on-board portable emissions monitoring system.	127

Figure 5.2 Inspiration gathered by the Professor from the experience taken in a summer school to an Assessment Activity for the students. 128
Figure 5.3 a) and b) Seminars with mobility experts (by videoconference and on-site); c) demonstration of hybrid and electric technology. 129
Figure 5.4 ERASMUS+ mission partnership between the Professors from Italy and Portugal: a) lecturing noise concepts and measurements methods; b) noise levels experimental monitoring in the university campus using Noise Capture app. 130
Figure 5.5 Visit to an automotive company. 130
Figure 5.6 Participation of bachelor and Master students in research activities (such as experimental monitoring campaigns). 135
Figure 5.7 Examples of traffic simulation (VISSIM software by PTV) and practical activities with an instrumented bicycle applied on Master dissertations. 135
Figure 5.8 Participation of Master students at international conferences on mobility and transportation, where they have the possibility to interact with senior researchers: example of the prestigious Transportation Research Board Annual Meeting, (Washington DC). 136
Figure 6.1 History of Education for Sustainable Development. 145
Figure 6.2 Highest Paying Industries for Environmental Engineers, U.S. 149
Figure 6.3 Major focus areas of Environmental Engineering. . 151
Figure 6.4 The 3C's Summarized add. 152
Figure 6.5 Grand Challenges of 21st Century. 153
Figure 6.6 Key Skills and Competencies for a 21st Century Engineer. 154
Figure 6.7 Key Contributions of Education for Sustainable Development. 155

List of Tables

Table 2.1 SDGs though the course 'Development and Growth' 33
Table 2.2 Dimensions of sustainable development and students' project topics 34
Table 3.1 Portuguese HEIs using STARS 56
Table 3.2 Measurement Method of the THE Impact Rankings to assess HEIs 57
Table 3.3 Portuguese HEIs in the THE Impact Rankings 59
Table 3.4 Best SDG Scores by Portuguese HEIs in the THE Impact Rankings 60
Table 3.5 Portuguese HEIs in the GreenMetric Rankings 63
Table 4.1 Evolution of the gross enrollment rate in higher education 90
Table 6.1 Key Events in the History of Sustainable Development 143
Table 6.2 Distribution of employments in different industries . . 149
Table 6.3 Key Sustainability Competencies and Guiding Principles for Engineers 157

List of Contributors

Bielschowsky, Carlos E., *Universidade Federal do Rio de Janeiro, Centro de Tecnologia, Instituto de Química, Cidade Universitária, Rio de Janeiro, RJ 21941-909, Brasil*

Boohene, David, *School of Management Sciences & Law, University of Energy and Natural Resources, Ghana; E-mail: david.boohene@uenr.edu.gh*

Caria, Ana, *Department of Management, School of Economics and Management, University of Minho, Campus Gualtar, 4710-057 Braga, Portugal; E-mail: aalexandra@eeg.uminho.pt*

Coelho, Margarida C., *Department of Mechanical Engineering, Centre for Mechanical Technology and Automation, University of Aveiro, Campus Santiago, 3810-193 Aveiro, Portugal; E-mail: margarida.coelho@ua.pt*

Correia, Paulo R. M., *University of São Paulo, São Paulo, Brasil; E-mail: prmc@usp.br*

Davim, J. Paulo, *Department of Mechanical Engineering, University of Aveiro, Campus Santiago, 3810-193 Aveiro, Portugal; E-mail: pdavim@ua.pt*

Franchetti, Matthew J., *The University of Toledo, 2801 W. Bancroft St., Toledo, OH 43606, USA; E-mail: matthew.franchetti@utoledo.edu*

Fu, H., *Northeast Petroleum University, Daqing, China; E-mail: fhl@nepu.edu.cn*

Hasan, Sibte, *Independent Researcher, Pakistan*

Hysa, E., *Epoka University, Tirana, Albania; E-mail: ehysa@epoka.edu.al*

Kalsoom, Qudsia, *Beaconhouse National University, Pakistan; E-mail: qudsia.kalsoom@bnu.edu.pk*

Kinchin, Ian M., *University of Surrey, Surrey, UK; E-mail: i.kinchin@surrey.ac.uk*

Kuruppuarachchi, Lakshika N., *The University of Toledo, 2801 W. Bancroft St., Toledo, OH 43606, USA; E-mail: lakshika.kuruppuarachchi@utoledo.edu*

Leal, Cristiana, *Department of Management, School of Economics and Management, University of Minho, Campus Gualtar, 4710-057 Braga, Portugal; Centre for Research in Economics and Management (NIPE), University of Minho; E-mail: ccerqueira@eeg.uminho.pt*

Machado, Carolina, *Department of Management, School of Economics and Management, University of Minho, Campus Gualtar, 4710-057 Braga, Portugal; E-mail: carolina@eeg.uminho.pt*

Mello, Sidney L. M., *Universidade Federal Fluminense, Escola de Engenharia, Laboratório de Tecnologia e Gestão de Negócios, Niterói, RJ, 24210-240, Brasil; Faculdade Cesgranrio, Rio de Janeiro, RJ, 22241-125, Brasil; E-mail: smello@id.uff.br*

Meriño, Marcelo J., *Universidade Federal Fluminense, Escola de Engenharia, Laboratório de Tecnologia e Gestão de Negócios, Niterói, RJ 24210-240, Brasil*

Odame, Christabel, *Department of Management Studies, Indian Institute of Technology-Dhanbad, India; E-mail: codame@anuc.edu.gh*

Oliveira, Benilde, *Department of Management, School of Economics and Management, University of Minho, Campus Gualtar, 4710-057 Braga, Portugal; E-mail: benilde@eeg.uminho.pt*

Oliveira, Lídia, *Department of Management, School of Economics and Management, University of Minho, Campus Gualtar, 4710-057 Braga, Portugal; E-mail: lidiaoliv@eeg.uminho.pt*

Panait, Mirela, *Petroleum-Gas University of Ploiesti, Ploiesti, Romania; Institute of National Economy, Bucharest, Romania; E-mail: mirela.matei@upg-ploiesti.ro*

Pandey, Mrinalini, *Department of Management Studies, Indian Institute of Technology-Dhanbad, India; E-mail: mrinalini@iitism.ac.in*

Petrescu, M. G., *Petroleum-Gas University of Ploiesti, Ploiesti, Romania; E-mail: pmarius@upg-ploiesti.ro*

Sampaio, Thaís N. da R., *Universidade Federal Fluminense, Escola de Engenharia, Laboratório de Tecnologia e Gestão de Negócios, Niterói, RJ 24210-240, Brasil*

Shaik, Salma, *The University of Toledo, 2801 W. Bancroft St., Toledo, OH 43606, USA; E-mail: salma.shaik@utoledo.edu*

List of Abbreviations

AASHE	Association for the Advancement of Sustainability in Higher Education
EfS	Energy for Sustainability
ESD	Education for Sustainable Development
FEUP	Faculty of Engineering of the University of Porto
GRI	Global Reporting Initiative
HEIs	Higher Education Institutions
IIRC	The International Integrated Reporting Council
ISCTE-UIL	Instituto Superior de Ciências do Trabalho e da Empresa – University Institute of Lisbon
MDGs	Millennium Development Goals
SCC 2019	Sustainable Campus Conference
SD	Sustainable development
SDGs	Sustainable Development Goals
STARS	Sustainability Tracking, Assessment and Rating System
THE	Times Higher Education
UC	University of Coimbra
UMinho	University of Minho
UN	United Nations
UN DESD	United Nations Decade of Education for Sustainable Development 2005–2014

1

Pedagogic Resonance and Threshold Concepts to Access the Hidden Complexity of Education for Sustainability

Paulo R. M. Correia[1,*] and Ian M. Kinchin[2]

[1]University of São Paulo, Brasil
[2]University of Surrey, UK
E-mail: prmc@usp.br; i.kinchin@surrey.ac.uk
*Corresponding Author

Abstract

Within this chapter, we conceptualize education for sustainability (EfS) as a wicked, socio-ecological problem that requires the implementation of a radical pedagogical perspective if it is to be addressed in a systemic and sustainable manner. The consideration of EfS through a complex, socio-ecological lens prompts a consideration from multiple perspectives that we articulate here as opposing sides of an epistemological abyss (or two-culture valley) – where rational, objective thinking (scientific reasoning) needs to interact with more personal, subjective knowledge structures. This allows us to approach key threshold concepts related to EfS that we identify as 'virtue ethics', 'wicked problems' and 'academic literacy'. Through the analogy of exploring unfamiliar terrain, we consider the importance of pedagogy as a foundational idea in the construction of interdisciplinary approaches to EfS in the classroom and the necessity of teachers acting as well-equipped guides who are familiar with both sides of the epistemological abyss.

Keywords: Pedagogic resonance, Threshold concepts, Knowledge structures, Wicked problems, Sophisticated instructional discourse.

1.1 Mind the Gap Between Pedagogy and Teaching Methods

The challenges of EfS are of increasing concern in Higher Education. The most frequent teaching practices in learning environments were configured in a different social context, to address the needs posed by the industrial society. A break with such practices is required because contemporary society faces increasingly complex problems. Openness to dialogue, critical thinking, consideration of ethical values, creativity and the ability to solve problems must be nurtured from new pedagogical architectures that include, for example, active learning methodologies and acknowledge increasing connectivity between all aspects of academic life.

The complexity of EfS has been discussed by several authors. The understanding of pedagogy and its role to frame innovative teaching methods has deserved special attention [1–3]. Some attempts to change teaching methods occur based only on the teachers' perception, without any deeper pedagogical foundation. In this case, the changes are based on the experiences teachers had as students or on passing fads and fashions that seem to be short-term solutions to renew the traditional classroom environment.

Several authors have highlighted the role of pedagogy in EfS. Lehtonen et al. claim the need for new learning approaches that enhance the critical reflection of dualistic and segregated ways of thinking and promote the awareness of interconnectedness [1]. According to them:

> The pedagogy of interconnectedness aims at enhancing the understanding of the world and humans as relational: recognizing the interdependence of society and nature, the local and global, and seeing the common reality as socially constructed. [1, p. 864]

The growing acknowledgement of the interconnectedness of complex systems (including education) needs to be aligned with systems thinking. Capra [4, p. 20–21] has identified a number of shifts in perception that are needed to accompany the adoption of systems thinking. These are summarized as shifts from the parts to the whole; from objects to relationships; from objective knowledge to contextual knowledge; from quantity to quality; from measurement to mapping; from structures to process, and from contents to patterns. This systemic approach offers a challenge to the typical fragmentation of the academic literature [5], and requires the reader to consider concepts and language that lie outside the normal academic silos of disciplinary discourses [6]. Bringing together the scientific, rational thinking of the quantitative disciplines into the same discussion as the complementary structures of personal experience (and indigenous beliefs) requires a level of post-abyssal thinking [*sensu* Santos, 7] in which the absolute epistemological sovereignty of science is relinquished to allow the voices of less powerful groups to be heard. Post-abyssal thinking can thus be summarized as learning from the 'other' side [of the abyssal line] through parallel epistemologies. It confronts the monoculture of modern scientific thought that dominates western universities, with the rich ecology of knowledge [described by Santos, 7]. Such epistemological pluralism is seen as the way forward for interdisciplinary research [e.g., 8, 9] - including that which investigates the processes that maintain complex socio-ecological systems (including education), where plurality can lead to a more sustainable and integrated consideration [10].

Mintz and Tal reported the benefits of using active and participatory teaching methods to foster the development of learning outcomes related to sustainability [2]. In a recent article, Sandri presents arguments for why explicit reflection on the meaning and the role of pedagogy is important for researchers and teachers [3]. She compares educational practice with the structure of an iceberg, in which the teaching methods and the educational approach are the observable part of the ice, whereas pedagogy is the underpinning block of ice; massive and unnoticed at the bottom of the structure:

> Applying the simple iceberg metaphor to educational practice can help demonstrate how pedagogy makes up the majority of the metaphorical iceberg that sits below the waterline and is not physically observable while the approaches (delivery, assessment style, content framing, learning objectives) and methods (activities, assessments, interactions, case studies) that are informed by pedagogy are more observable forms of educational practice [3, p. 5].

The connection between teaching methods and pedagogy occurs through an intermediate layer, which deserves to be further clarified. We argue the educational approach plays a critical role to bridge pedagogy (the synthetic expression of the regulative discourse) and teaching methods (the practical expression of the instructional discourse). For this reason, we modified and redrew the original iceberg (Figure 1.1(a)) to make it even more representative: only the teaching methods are at the visible tip, while the invisible portion is bigger than originally suggested.

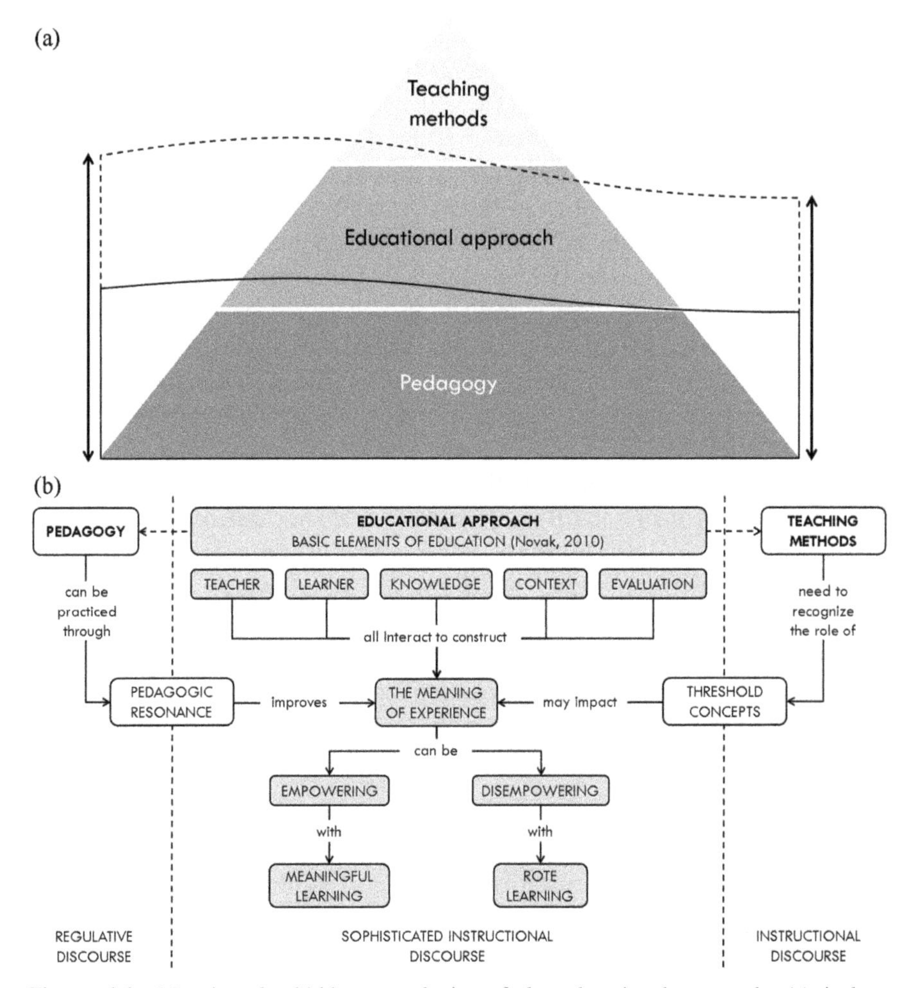

Figure 1.1 Mapping the hidden complexity of the educational approach: (a) iceberg metaphor to highlight that only the tip is visible to external observation (redrawn and modified from [3] and (b) the sophisticated instructional discourse framed from Novak's basic elements of any educational event grey boxes, redrawn and modified from [11]).

The five elements of educational events proposed by Novak [11] bring the 'Didactic realm' to reveal some hidden aspects of the 'educational approach'. The meaning of experience results from the interaction of teacher, learner, knowledge, content and evaluation, which can result in empowering (meaningful learning) or disempowering (rote learning) outcomes. Figure 1.1(b) highlights the meaning of experience as the central concept in the hidden complexity of the educational approach layer. Moreover, it provides connections with pedagogy (bottom layer) and teaching methods (upper layer) through pedagogic resonance and threshold concepts.

Our chapter explores the role of pedagogic resonance [12] and threshold concepts [13] to highlight the hidden complexity in EfS, through a sophisticated instructional discourse that considers the meaning of the experience and the nature of the knowledge involved. We discuss learning as changes in knowledge structures to create new meanings, challenging learners to struggle with critical moments when the content seems to be meaningless (disjuncture), and periods when no relevant gain of understanding occurs (conceptual stasis).

1.2 Learning as a Climbing Adventure

Learning is not the calm and peaceful process it seems, particularly if we choose to learn meaningfully. The creation of new meaning by combining previous knowledge and new information is an adventure marked by points of disjuncture and leaps. We can compare the learning process to an expedition to conquer a mountain range with various obstacles to be overcome. The obstacles are imposed by the conceptual landscape of the terrain, which is seen by the learners and unnoticed by the experts who tend to smooth out the lumps and bumps in their learning and forget what it is to be a novice [14]. For this reason, mapping the terrain is recommended for mountain guides who wish to scaffold the first-time explorers through the conceptual mountain range of EfS.

Knowledge structures change throughout the learning process. The initial perception that conceptual changes are gradual and occur all the time is not compatible with reality. The examination of our experience is sufficient to recognize that there are remarkable moments when, after a long effort, we finally understand the yet-to-be-known. From that moment on, there is a leap in the meanings we can produce from the subject under study. Unfortunately, we do not jump all the time while learning because conceptual leaps require preparation, during which nothing seems to happen. The punctuated learning model [15] values conceptual stasis as the preparatory moment for the construction of meanings, which marks the leaps in understanding that usually occur when we cross threshold concepts. As Kinchin highlights:

> Stasis is required as part of the learning process: 'lining up' the segmental and cumulative knowledge structures for subsequent

> integration. [...] The thresholds create moments of transformative change whilst the periods of conceptual stasis, rather than being 'nothing', are required to assemble the raw materials that will facilitate that change [15 Kinchin, 2010, p. 56–57].

1.2.1 Disjuncture and The Need for Pedagogic Resonance

The concept map presented in Figure 1.2(a) compares three knowledge structures that have been revealed by concept mapping studies into student understanding [16]. Spoke and network knowledge structures represent understanding according to its richness. Novices have limited and superficial knowledge about a central topic, which is represented by spoke structures. On the other hand, specialists can articulate more concepts on the subject and this in-depth knowledge produces networks in which the interconnections between concepts confers greater understanding. Linear structures are related to procedural knowledge that is usually goal-oriented. Integrations involving understanding (conceptual knowledge) and practice (procedural knowledge) occur through learning leaps that are critical to building experts' knowledge. They can contextualize the procedural knowledge and select the best chains of practice to fit the specific needs of the problem at hand [15, 17]. Conversely, beginners face difficulties in establishing relationships between theory and practice. In this situation, memorization of concepts and procedures becomes a potential undesirable outcome. Pedagogical resonance is the antidote to avoid rote learning. It creates an authentic space for dialogue between experts and novices, who act as epistemological agents during a process of collaborative meaning-making.

Trigwell and Shale [12] stress the importance of collaborative meaning-making and the dynamic nature of engagement with students:

> It is the quality of awareness that is evoked in collaborative meaning-making with students that define the quality of a teacher's response to the teaching situation. It is this evoked awareness – the dynamic, reciprocal, fluid engagement with students – and related action that we must seek to capture if we are to truly represent student-focused teaching in an analysis of the scholarship of teaching. This evoked or relational awareness/action is what we call pedagogic resonance. [12, p. 523]

Hay et al. [18] show that linear structures are limited in representing conceptual understanding. Its initial growth soon becomes an obstacle to the

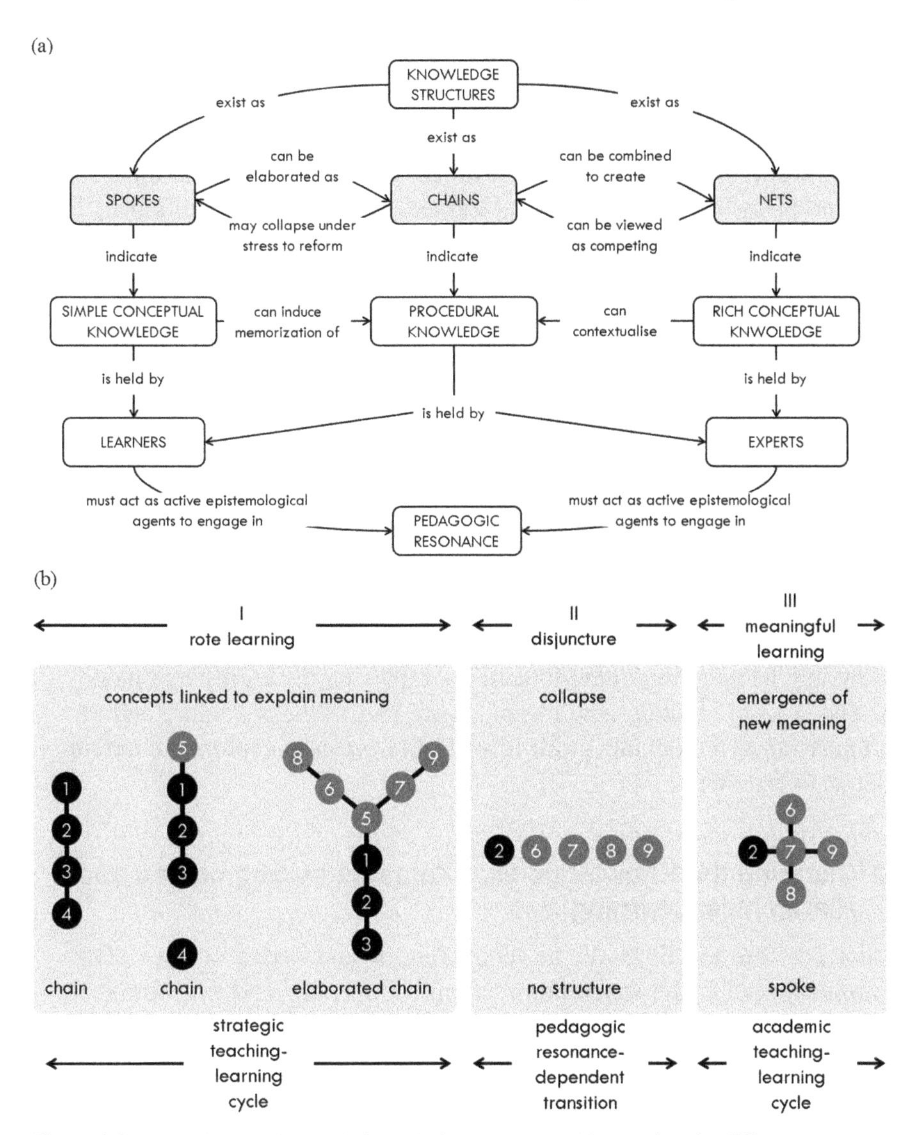

Figure 1.2 Learning as changes in knowledge structures: (a) mapping the differences among spokes, chains and nets (redrawn and modified from [16]) and (b) the disjuncture as a daunting event to avoid rote learning (redrawn and modified from [18]).

inclusion of more information (Figure 1.2(b)). At this point, beginners are faced with a choice that can lead to rote (memorizing the additional information) or meaningful learning (requiring a reorganization of prior knowledge).

The latter involves a drastic transformation of the knowledge structure to enable the construction of a new hierarchy of meanings. This change is described by Hay and colleagues as follows:

> The student in this case study began a course of learning with a simple prior-knowledge structure and learned, at first, by rote addition. Later, however, they found that what was new was irreconcilable with what they had understood to begin with. The result was a period of 'disjuncture', during which the student was less able to explain the topic than they had been before. Eventually, they achieved a new grasp of meaning, but this came after a difficult period in which they might easily have given up. [18, p. 304]

The period of disjuncture (Figure 1.2(b)) is marked by the rupture of the linearity and the emergence of a spoke structure, more appropriate for the organization of conceptual knowledge (Figure 1.2(a)). The emergence of new meanings only occurs after overcoming the disjuncture, which is the period when learners cannot internalize or integrate new information.

The daunting experience, generated by repeated periods of disjuncture, is faced by any learner committed to actively exploring the knowledge mountains. The risks of creating new meanings are part of the adventure and the expert must act as a mountain guide to scaffold novices to explore the terrain that is yet to be known.

1.2.2 Climbing the Knowledge Mountains Using Segmented and Hierarchical Learning

It is not possible to climb all the mountains of knowledge at once. This will certainly lead to the adventurers' cognitive overload and therefore it is necessary to map the terrain to establish the best locations for the base camps. These are resting places for the adventurers getting ready for the next climb, *i.e.*, for the leap of understanding that will be achieved during the exploration of the yet-to-be known (Figure 1.3).

This analogy is useful to understand the intercalation of conceptual stasis and conceptual changes. Segmented and cumulative learning [*sensu* Maton, 19] takes place during conceptual stasis [15]. They are related to linear and hierarchical structures, respectively. In the base camps (Figure 1.3), learners organize the knowledge needed for the next climb. It is interesting to note that theoretical and practical aspects are refined during this period, through

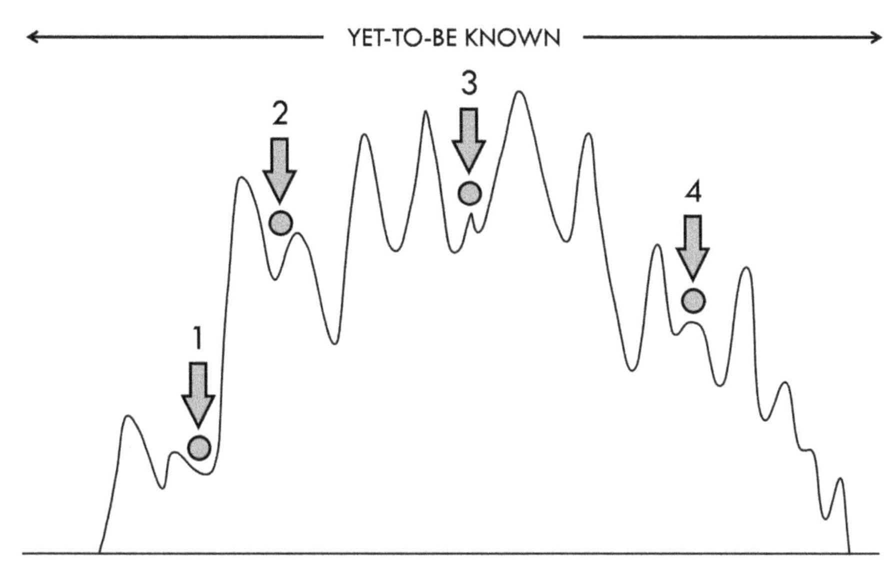

Figure 1.3 Develop segmented and hierarchical knowledge structures is a strategy to conquer the knowledge mountains. The arrows 1–4 indicate base camps that serve as preparation (conceptual stasis) before climbing the highest peaks (conceptual change).

activities that the expert proposes to get the learners ready for the leap of understanding. The identification and remediation of alternative conceptions tend to take place at this point. This process involves the conceptual reconstruction of structures that are limited or inappropriate propositional hierarchies (LIPH). As they were internalized by students, meaningful learning is the unique option the learners have to remedy them [20]. According to Novak:

> Only the learner can choose to do this, so one of the obstacles to alterations of misconceptions or LIPH's is that learners who choose to learn by rote will not modify their existing knowledge structures regardless of the efforts of the text or instructor. Thus a precondition for remediating a given LIPH is that learners must choose meaningful learning, at least to some degree. Furthermore, high levels of meaningful learning require that the learner already possess a relatively sophisticated relevant knowledge structure, so in most cases, remediation of LIPHs will be an iterative process where the learner gradually builds relevant knowledge structures and refines these over time. [20, p. 558]

1.3 Mapping the Conceptual Terrain

Knowing the terrain is an essential aspect of the expedition's success. Identifying where the highest peaks are and defining the locations to establish the base camps are crucial aspects to be considered. The experts have this responsibility [see 21], and they must act as experienced guides anticipating the challenges that lie ahead. The plan for the adventure must recognize the periods of disjunction and prepare students to keep their commitment to learning meaningfully throughout the way.

The knowledge accumulated by disciplinary experts tends to produce a simplified view of the path to be taken by learners. Teachers can acknowledge this difference and some academics employ geographical or topographical analogies to articulate this, seeing themselves as guides for their students. For example:

> Learning is like a waterfall. I know there's a path up the back to the top so students can dive off, but someone needs to take them up the path. But to me, that's exactly it. I'm there to signpost. I'm there to go 'come on, this way, this way'. I'm not there to dictate what that journey looks like. But I'm there to show the way of the journey and say 'come on'. [22, p. 102]

Differences between the novice and expert perceptions of learning challenge is problematic, as the preparation steps (during periods of conceptual stasis) may be insufficient to overcome the higher mountains (conceptual change). Therefore, route changes during the expedition may be necessary even when the initial plan is well devised. The adaptability of experts and novices must be at the service of meaningful learning, empowering all the participants (Figure 1.1(a)). Pedagogic resonance is the key element for implementing the adaptations that will guarantee the success of climbing the mountain range.

1.3.1 Threshold Concepts and Changes in Understanding

It is evident that not all concepts are equivalent when we build a hierarchy of meanings. A few are central to the organization, while many others are peripheral. Conceptual understanding increases as knowledge structures become stable and well-organized. Several transformations occur during the construction and revision of these structures during periods of conceptual stasis. On the other hand, a few concepts are capable of producing dramatic transformations in our understanding of the topic under study. These are remarkable moments that bring us closer to constructing expert knowledge. Threshold concepts are responsible for the most significant conceptual changes. They are the highest mountains that appear throughout our expedition (Figure 1.3). The role of threshold concepts in understanding is described by Meyer and Land as follows:

> A threshold concept represents a transformed way of understanding or interpreting or viewing something without which the learner cannot progress. As a consequence of comprehending a threshold concept there may thus be a transformed internal view of subject matter, subject landscape or even world view. [...] Such a transformed view or landscape may represent how people 'think' in a particular discipline or how they perceive, apprehend or experience particular phenomena within that discipline (or more generally). [13, p. 3]

Meyer and Land offer key characteristics of threshold concepts that distinguish them from other important ideas within a discipline. Threshold concepts are likely to be:

1. Transformative: they result in a change in perception of a subject and may involve a shift in values or attitudes.
2. Irreversible: the resulting change is unlikely to be forgotten.
3. Integrative: it 'exposes a previously hidden interrelatedness' of other concepts within the discipline.
4. Bounded: it serves to define disciplinary areas or to 'define academic territories'.
5. Potentially troublesome: students may have difficulty coping with the new perspective that is offered.

The difficulty in acquiring threshold concepts may leave the learner in a state of liminality, a suspended state in which understanding approximates a kind of mimicry or lack authenticity [13] or where understanding appears to collapse (Figure 1.2(b)). Meyer and Land also develop the argument that acquiring a threshold concept may be linked in some disciplines to a 'rite of passage' when novices become closer to the experts:

> The term 'liminality' (from the Latin *limen*, boundary or threshold) was also adopted by Turner to characterise the transitional state of space or time within which rituals are conducted. It should come as no surprise that this notion of a 'rite of passage' resonates strongly in many disciplines with entry into their communities of practice. [13, p. 22]

It is interesting to note that students go again through a period of a sudden loss of understanding. The disjuncture, which may occur during conceptual stasis, is similar to the liminality that happens during the profound conceptual change caused by acquiring threshold concepts. Comparing learning with a climbing adventure gives a more realistic way of representing the challenges one faces when trying to construct knowledge in a non-arbitrary way. During this adventure, teachers need to be aware of the importance of pedagogical resonance as a space for negotiating meanings with students, mainly when they are disoriented due to disjuncture or liminality. Paraphrasing Novak, who proposes 'meaningful learning as the essential factor for conceptual change' [20] we affirm that 'pedagogical resonance is the essential factor for meaningful learning'.

Climbing the mountains allows one to get a different sight of the broad landscape. This is a reward after all efforts to reach the knowledge peaks that can transform us. Similarly, the transformative effect of crossing a threshold concept changes the way we perceive the conceptual terrain. The initial mountains became a flat terrain (Figure 1.4) because the past learning obstacles are not noticed after crossing threshold concepts. However, they are still there in the conceptual terrain, waiting for more novices.

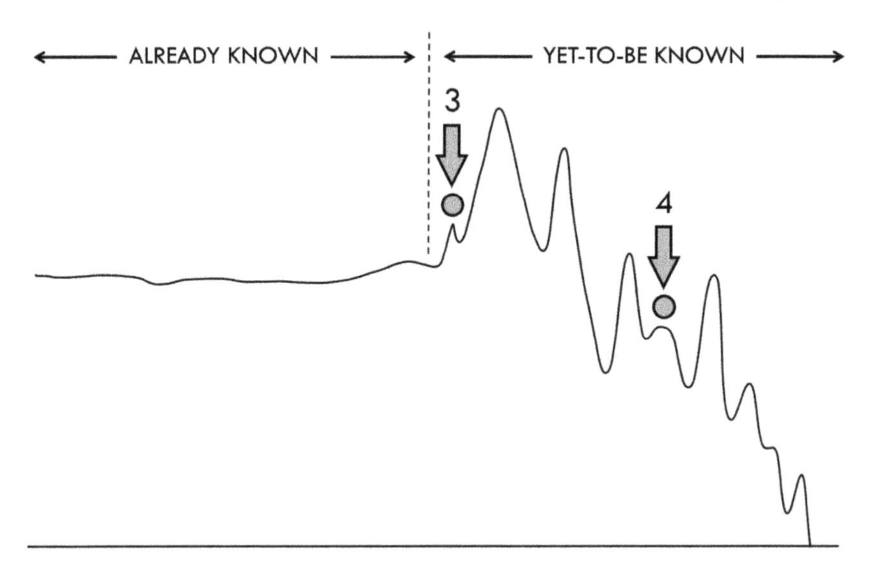

Figure 1.4 Overcoming threshold concepts flatten the landscape, and experts struggle to understand the learners' difficulties with the apparent flat terrain (already known). The base camps for the next climbing (arrows 3–4) are visible because they help to enter the 'yet-to-be known'.

This situation poses an additional challenge to teachers who forget what it is like to be a novice [14]. The consequence is that classes are offered beyond the capabilities of those who climb the mountains for the first time, because of the forgetfulness of the conceptual complexity of the terrain. The identification of the main learning obstacles helps the teacher to select suitable content and teaching methods at each moment of the learning process. This warning is even more relevant in the interdisciplinary context of EfS, where the conceptual complexity increases due to the nature of the knowledge involved.

1.3.2 Knowledge Integrations at The Threshold Concepts

The changes in understanding that are associated with threshold concepts can be explained by the integration of knowledge structures. This process increases the possibilities of producing new meanings, through the connection between knowledge structures that were not previously related. Threshold concepts can be understood as these connection points, which expand the conceptual territory available for future learning.

Figure 1.5 presents different integrations between knowledge structures that can be mediated by threshold concepts. The first example represents

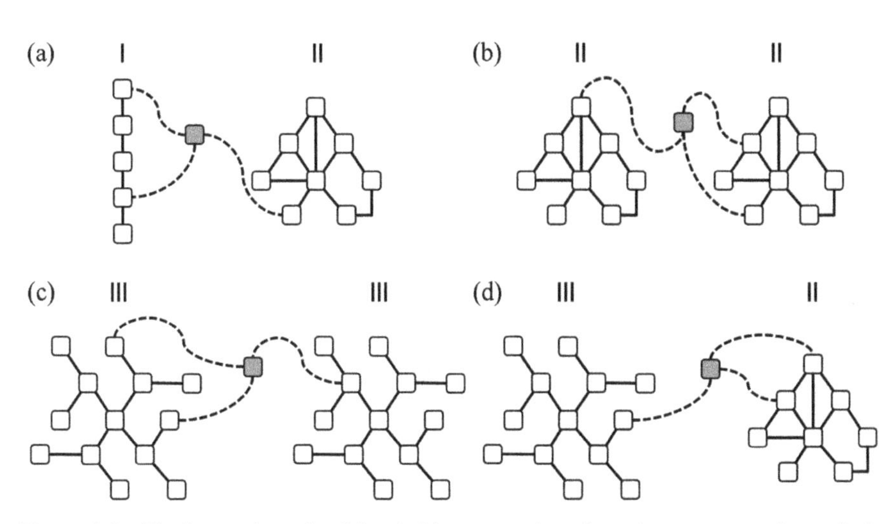

Figure 1.5 The integrative role of threshold concepts (grey boxes) to overcome the peaks in the knowledge mountain range. Four different integrations are shown considering procedural knowledge (I), upper-tree hierarchy (conceptual knowledge from scientific disciplines, II) and network hierarchy (conceptual knowledge from humanities, III).

the approximation between a linear structure (procedural knowledge) and a hierarchical structure. The latter refers to the knowledge structure typical of scientific disciplines [23], and is often associated with professional knowledge where the threshold concept facilitates the essential connection between theory and practice (Figure 1.5(a)), contextualizing the chains of practice from conceptual understanding [15, 17, 24].

Other integrations are also noteworthy, especially when they broaden conceptual horizons. Figure 1.5(b) shows the integration between two hierarchical knowledge structures, representing the deepening of understanding of scientific topics. It is worth remembering that Donald showed that scientific knowledge is marked by a hierarchical organization [23], justifying the tree-like structure. This representation becomes even more interesting when we consider that there is more than one discipline involved in integration. Biochemistry (Biology and Chemistry) and Astrophysics (Astronomy and Physics) are interdisciplinary efforts that broaden our understanding. In other words, interdisciplinarity is a way to expand the terrain and explore mountains that were not seen at the beginning of the adventure. Interdisciplinary fields are not simply additive but can create synergies that offer new conceptual lenses to view the terrain and new concepts to help articulate what is being viewed.

A similar situation can be seen within the humanities disciplines. In this case, Figure 1.5(c) represents the approximation of network structures, which represent the organization of knowledge in these disciplines [23]. The disciplinary approach of this integration can be exemplified from the deepening of knowledge related to Education, Sociology or Psychology. On the other hand, the integration of these structures will be interdisciplinary if the knowledge is related to Educational Sociology or Educational Psychology.

Figure 1.5(d) represents a very common situation when the contents of EfS are discussed. In this case, the integration of structures is interdisciplinary and involves scientific disciplines and the humanities. Snow [25] was already aware of the existence of two academic cultures with serious difficulties in establishing productive communication. However, the complex challenges of contemporary society demand the approximation of sciences and humanities to expand the conceptual field available to seek the best possible solutions. Wilson [26] has described this as the 'jumping together' or 'consilience' of 'the great branches of learning'. Scientific literacy, for example, is an indispensable concept for understanding topics related to EfS. Correia et al., [27] pointed out that the interdisciplinary integration necessary for exploring scientific literacy was more complex than the cases in which knowledge is from a single epistemological viewpoint – i.e. from the objective, scientific perspective *or* from the more subjective arts/humanities:

> Scientific literacy objectives are achieved only when techno-scientific and humanistic perspectives are simultaneously and evenly considered, allowing educational communities to overcome the gap that isolates the 'two cultures'. [27, p. 680]

EfS is a prime example demonstrating the need for an epistemologically plural perspective, e.g.:

> Becoming (consciously) bi- or multi-epistemic or operational in two or more ways of knowing, involves understanding different social and historical dynamic processes of knowledge construction, their limitations and the social-historical relations of power that permeate knowledge production. It also involves being able to reference, combine and apply the appropriate frame of reference to an appropriate context. [8, p. 46]

This suggests that epistemology is something that needs to be brought to the fore within the EfS classroom, as well as in teacher development programmes, in order to develop skills and tools to help traverse the epistemological abyss as a matter of routine [e.g. 28–30].

1.4 Charting the EfS Knowledge Mountains

This section invites the reader to join us in the effort to map the conceptual terrain of EfS. Figure 1.6 is a preliminary version containing some high peaks that stand out in the landscape. They form a nested hierarchy of threshold concepts. Although incomplete, this hierarchy serves as a starting point for detailing the map that will indicate the location of the main learning obstacles. From them, the expert can define the safest directions for climbing the knowledge mountains to ensure that most learners complete the route.

Our hierarchy starts recognizing sustainability as the broadest threshold concept. The space-time materialities, scale impasse and the dilemmas that emerge from cultural and social structures need to be combined to grasp the complexity of the EfS challenges. According to Murphy [31]:

> Three difficulties are particularly important in making sustainability a wicked problem. The first consists of the distant consequences of unsustainable practices in both time and space. The experience

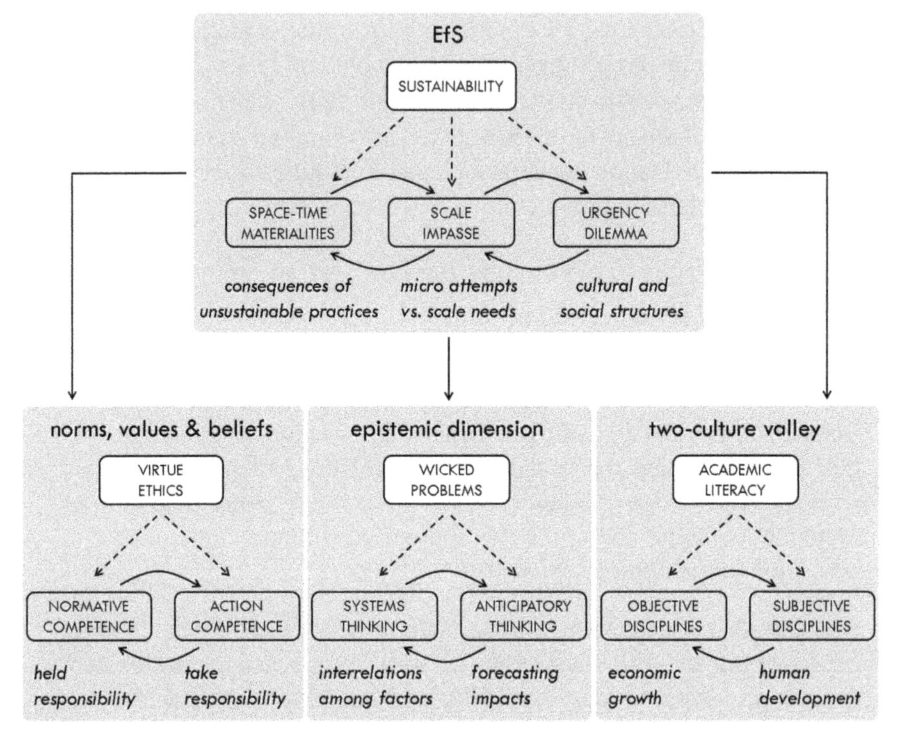

Figure 1.6 A preliminary nested hierarchy of threshold concepts (white boxes) to access the hidden complexity of EfS. Grey boxes detail the relevance of 'virtue ethics', 'wicked problems' and 'academic literacy' to understand 'sustainability'.

> of local well-being in the present seems to contradict scientific indications of threats for the future, hence it is easy for local groups to ignore the warnings and to continue the habitus. The second is the scale impasse, namely that successful local micro attempts at sustainability run into difficulty when tried on the mass scale necessary for sustainability. The third is the urgency dilemma. The threats to sustainability are embedded in cultural and social structures, habitus and the built physical infrastructures such that it will take time to change them, but those threats require a prompt change of those structures. [31, p. 24]

We select three subordinated threshold concepts to detail the hidden complexity of sustainability. Blok and colleagues [32] nicely frame the ethical dimension proposing the virtue ethics to contrast the normative competence (held responsibility) and the action competence (take responsibility):

> Virtuous competence can be defined as the personal engagement of a professional in the transformation to a good character by applying virtues in the production of sustainable internal goods (corporate sustainable behaviour in collaboration with and in response to multiple stakeholders, and by perfecting his or her good character by internalisation of the production of these goods. [32, p. 23]

Wicked problems are frequently present in discussions involving sustainability issues [1, 31–34]. Despite the lack of a consistent application of the term pointed out by Lönngren and van Poeck [34], it is a necessary part of the knowledge hierarchy in Figure 1.6 to highlight the role of the epistemic dimension. Blok and colleagues [32] argue that:

> Wicked problems are highly complex because they concern global issues like climate change, desertification and poverty and cannot be solved in traditional ways or by simple solutions; the complexity of sustainability consists in the fact that cause and effect relations are either unknown or uncertain and that multiple stakeholders are involved with differing ideas about what the 'real' problem is and often having conflicting norms, value frames and beliefs regarding the subject. [32, p. 5]

Systems thinking [4, 33] and anticipatory thinking [33] are critical skills needed to develop to respond to wicked problems. Lans and colleagues [33] clarify the characteristics of systems and anticipatory thinking as follows:

> With regard to sustainable development, systems-thinking is the ability to identify and analyse all relevant (sub)systems across different domains (people, planet, profit) and disciplines, including their boundaries. Furthermore, systems-thinking is the ability to understand and reflect upon the interdependency of these (sub) systems including cascading effects, inertia, and feedback loops and accompanying cultures. [...] The ability to collectively analyse, evaluate and craft 'pictures' of the future in which the impact of local and/or short-term decisions on environmental, social and economic issues is appreciated on the global/cosmopolitan scale and the longer term. This capacity includes skills in creativity, opportunity recognition, innovation and balancing of local/global and short-term/long-term perspectives. [33, p. 40]

Finally, we acknowledge the importance of recognizing to pursue academic literacy as a way to overcome the valley that isolates objective and subjective disciplines. As Correia [35] pointed out:

> 'Two cultures' is the term coined by C. P. Snow (1959) in his legendary lecture at the University of Cambridge. In 1959, Snow already noticed the differences between Natural Sciences and Humanities. In a nutshell, Snow condemned literary scholars for unfamiliarity with the Second Law of Thermodynamics - the scientific equivalent of knowing Shakespeare's work, lamenting the chasm between intellectuals and scientists, as well as the distorted image that a group had of other. The expression 'two cultures' suggests the distance that separates these worlds, aimed at the production of knowledge to unveil reality. Ontological differences arise from the selection of study objects, research paradigms, methods of investigation and the language typically used to communicate the achieved results. These differences explain the chasm that isolates the Natural Sciences and Humanities at the university. The knowledge specialization has produced a dark valley that needs to be overcome through interdisciplinary efforts that involve academics who barely recognize themselves as peers. [35, in press]

1.5 Implications for Mountain Guides and Adventurers

Understanding of the learning process and conceptual terrain is not fully appreciated when the EfS discussion is framed from Pedagogy and teaching methods. The sophisticated instructional discourse between them uncovers the hidden complexity at the 'educational approach' layer of the iceberg metaphor and the epistemological pluralism that needs to be embedded within it.

The option to undertake meaningful learning involves risk and can lead to periods of disjuncture (conceptual stasis) and a state of liminality prior to conceptual change. Pedagogic resonance is critical to avoid rote learning. Expert support can make a difference during the loss of meaning that occurs when novices are climbing the knowledge mountains, particularly when the guide has a knowledge of both sides of the epistemological abyss [7].

Conceptual stasis, which happens at the 'base camps', is the occasion where the identification and remediation of misconceptions take place through the revision of segmental and hierarchical knowledge structures.

These are important periods in the learning trajectory of the student. Despite external appearances of inactivity, these are not periods where nothing happens. Without these conceptual pauses, student anxiety is likely to result in a return to the familiarity of rote learning.

Leaps of understanding are integrations of knowledge structures that bring the novice closer to the expert knowledge structure. They occur when we climb the mountains and acquire threshold concepts. These knowledge integrations help to reveal that novices and experts perceive the learning obstacles differently. Integration can involve procedural and conceptual knowledge that need to be receptive to each other through repeated interaction [24], connecting theory and practice. Pedagogic resonance is required to support learners to overcome threshold concepts, minimizing the expert's oversimplification of the conceptual terrain.

Interdisciplinary adds complexity to the integration process because it involves conceptual knowledge structures from scientific disciplines (upper tree hierarchy) and humanities (network hierarchy). This requires the expert to develop a degree of epistemological flexibility to navigate the different knowledge structures and different academic cultures.

1.6 Acknowlegdements

P.R.M.C. thanks to São Paulo Research Foundation (FAPESP, grants #2016/24553–7 and #2012/22693–5), and National Council for Scientific and Technological Development (CNPq) for funding his research group.

References

[1] A. Lehtonen, A. Salonen, H. Cantell, L. Riuttanen, 'A pedagogy of interconnectedness for encountering climate change as a wicked sustainability problem', *Journal of Cleaner Production*, vol. 199, pp. 860–867, 2018.

[2] K. Mintz, T. Tal, 'The place of content and pedagogy in shaping sustainability learning outcomes in higher education', *Environmental Education Research*, vol. 24, no. 2, pp. 207–229, 2018 , DOI: 10.1080/13504622.2016.1204986

[3] O. Sandri, 'What do we mean by 'pedagogy' in sustainability education?', Teaching in Higher Education, 2020, DOI: 10.1080/13562517.2019.1699528

[4] F. Capra, 'Speaking nature's language: Principles for sustainability'. In: M. K. Stone & Z. Barlow (Eds.), Ecological literacy: Educating our children for a sustainable world (pp. 18–29), *San Francisco: Sierra Book Club Books*, pp. 18–29, 2005.
[5] D. Bohm, D. 'Wholeness and the implicate order', Abingdon, *Routledge*, 1980.
[6] I.M. Kinchin, K. Gravett, 'Dominant discourses in higher education: Critical perspectives, cartographies and practice', London: Bloomsbury, 2022.
[7] B.S. Santos, 'Epistemologies of the South: Justice against epistemicide', London: *Routledge*, 2014.
[8] V.D.O. Andreotti, C. Ahenakew, G. Cooper, 'Epistemological pluralism: Ethical and pedagogical challenges in higher education', AlterNative, vol. 7, no. 1, pp. 40–50, 2011. https://doi.org/10.1177/117718011100700104
[9] H. Suri, 'Epistemological pluralism in research synthesis methods', *International Journal of Qualitative Studies in Education*, vol. 26, no. 7, pp. 889–911, 2013. https://doi.org/10.1080/09518398.2012.691565
[10] T.R. Miller, T.D. Baird, C.M. Littlefield, G. Kofinas, F. Stuart Chapin III, C.L. Redman, 'Epistemological pluralism: reorganizing interdisciplinary research', *Ecology and Society*, vol. 13, no. 2, pp. 46, 2008. www.jstor.org/stable/26268006
[11] J.D. Novak, 'Learning, creating, and using knowledge: Concept maps as facilitative tools in schools and corporations', 2nd Edition, New York: *Routledge*, 2010.
[12] K. Trigwell, S. Shale. 'Student learning and the scholarship of university teaching', *Studies in Higher Education*, vol. 29, no. 4, pp. 523–536, 2004.
[13] J. Meyer, R. Land, 'Overcoming barriers to student understanding: Threshold concepts and troublesome knowledge', New York: *Routledge*, 2006.
[14] S.I. Fontaine, 'Teaching with the beginner's mind: Notes from my karate journal', College Composition and Communication, *Athens*, vol. 54, no. 2, pp. 208–221, 2002.
[15] I.M. Kinchin, 'Solving Cordelia's Dilemma: threshold concepts within a punctuated model of learning', *Journal of Biological Education*, vol. 44, no. 2, pp. 53–57, 2010.
[16] I.M. Kinchin, S. Lygo-Baker, D.B. Hay, 'Universities as centres of non-learning', *Studies in Higher Education*, vol. 33, no. 1, pp. 89–103, 2008.

[17] I.M. Kinchin, L.B. Cabot, M. Kobus, M. Woolford. 'Threshold concepts in dental education', *European Journal of Dental Education*, vol. 15, no. 4, pp. 210–215, 2011.
[18] D. Hay, I. Kinchin, S. Lygo-Baker, 'Making learning visible: the role of concept mapping in higher education', *Studies in Higher Education*, vol. 33, no. 3, pp. 295–311, 2008.
[19] K. Maton, 'Cumulative and segmented learning: Exploring the role of curriculum structures in knowledge-building', *British Journal of Sociology of Education*, vol. 30, no. 1, pp. 43–57, 2009.
[20] J.D. Novak, 'Meaningful learning: The essential factor for conceptual change in limited or inappropriate propositional hierarchies leading to empowerment of learners', *Science Education*, vol. 86, no. 4, pp. 548–571, 2002.
[21] C. Winch, 'Curriculum design and epistemic ascent', *Journal of Philosophy of Education*, vol. 47, no. 1, pp. 128–146, 2013.
[22] I.M. Kinchin, C. Derham, C. Foreman, A. McNamara, D. Querstret, 'Exploring the salutogenic university: Searching for the triple point for the becoming-caring-teacher through collaborative cartography' *Pedagogika*, vol. 141, no. 1, pp. 94–112, 2021. ISSN2029-0551_2021_V_141_1.PG_94-112.pdf
[23] J.G. Donald, 'Knowledge structures: Methods for exploring course content', *The Journal of Higher Education*, vol. 54,, no. 1, pp. 31–41, 1983.
[24] I.M. Kinchin, N.E. Winstone, E. Medland, 'Considering the concept of recipience in student learning from a modified Bernsteinian perspective', *Studies in Higher Education*, 2020. https://doi.org/10.1080/03075079.2020.1717459
[25] C.P. Snow, 'The two cultures and the scientific revolution', Cambridge: University Press; 1959.
[26] E.O. Wilson, 'Consilience: The unity of knowledge', London, Abacus, 1998.
[27] P.R.M. Correia et al., 'The importance of scientific literacy in fostering education for sustainability: Theoretical considerations and preliminary findings from a Brazilian experience', *Journal of Cleaner Production*, vol. 18, no. 7, pp. 678–685, 2010.
[28] E. Osborne, V. Anderson, B. Robson, 'Students as epistemological agents: claiming life experience as real knowledge in health professional education'. *Higher Education*, vol. 81, no. 4, pp. 741–756, 2021.

[29] J.C. Tovar-Gálvez, 'The epistemological bridge as a framework to guide teachers to design culturally inclusive practices', *International Journal of Science Education*, pp. 1–17, 2021, https://doi.org/10.1080/09500693.2021.1883203

[30] I.M. Kinchin, A.E. Thumser, 'Mapping the 'becoming-integrated-academic': An autoethnographic case study of professional becoming in the biosciences. *Journal of Biological Education*, 2021, https://doi.org/10.1080/00219266.2021.1941191

[31] R. Murphy, 'Sustainability: A Wicked Problem', Sociologica: *Italian Journal of Sociology on Line*, vol. 2, 2012, doi: 10.2383/38274

[32] V. Blok, B. Gremmen, R. Wesselink, 'Dealing with the wicked problem of sustainability', *Business and Professional Ethics Journal*, 2016, https://doi.org/10.5840/bpej201621737

[33] T. Lans, V. Blok, R. Wesselink, 'Learning apart and together: towards an integrated competence framework for sustainable entrepreneurship in higher education', *Journal of Cleaner Production*, vol. 62, pp. 37–47, 2014.

[34] J. Lönngren, K.v. Poeck, 'Wicked problems: a mapping review of the literature', *International Journal of Sustainable Development & World Ecology*, 2020, doi: 10.1080/13504509.2020.1859415

[35] P.R.M. Correia, 'Building a bridge from Chemistry to Education to overcome the valley between the two cultures'. In: N. Rao, A. Hosein, I.M. Kinchin (Eds.) 'Narratives of becoming leaders in disciplinary and institutional contexts: Leadership identity in learning and teaching in higher education', London, Bloomsbury, 2022, in press.

2

Universities – Players in the Race for Sustainable Development

M. Panait[1,2,*], E. Hysa[3], M. G. Petrescu[1], and H. Fu[4]

[1]Petroleum-Gas University of Ploieşti, 100680, Ploieşti, Romania
[2]Institute of National Economy, Bucharest, Romania
[3]Epoka University, Tirana, Albania
[4]Northeast Petroleum University, Daqing, China
E-mail: mirela.matei@upg-ploiesti.ro; ehysa@epoka.edu.al; pmarius@upg-ploiesti.ro; fhl@nepu.edu.cn
*Corresponding Author

Abstract

Promoting the principles of sustainable development supposes the involvement of many categories of stakeholders. Universities have become important players in the complex process of transition to the green economy. Their contribution in this process is particularly important considering the multiple functions of universities in didactic, research and entrepreneurial fields. Globalization, international crises, the pressures generated by the intensification of international competition, the aging of the population, the challenges generated by the technical progress are some of the factors that have determined the remodelling of the activities carried out by universities. The functions and role of universities have adapted to the development of society and these entities, given the spirit of innovation that characterizes them, have provided solutions to the challenges facing nations. This chapter aims to identify the main directions of action of universities in the process of promoting sustainable development. These directions are correlated with the functions that universities fulfil, taking into account the complexity of the sustainable development phenomenon and the different categories of stakeholders involved.

Keywords: Universities, research, entrepreneurship, sustainability

2.1 Introduction

We are thus witnessing the transformation of universities into economic agents and even leaders of economic growth and sustainable development, in a first phase for the region of which they are part, which ensures the transformation of knowledge into the own capital of academics. Universities are also trying to find solutions to the challenges posed by climate change and the transition to the green economy. For this reason, they are important actors in the process of promoting sustainable development both through their own actions and through the actions of the specialists they train [1, 2].

Sustainability is a concept that now reshapes the teaching, research and innovation activity of universities but also their societal role, higher education institutions becoming regional hubs in promoting the principles of sustainable development. Thus, universities have adhered to the Compact Global Principles launched by United Nations in 2000. This platform brings together different types of entities such as companies, unions, cities, public sector organisations, NGOs, business associations, universities. According to the statistics available on the site of Global Compact, almost 700 academic organisations and over 6.5000 companies adhered to these principles [3].

Universities, companies and other interested entities adopt these principles voluntarily. Despite the lack of mandatory regulations, a commitment to implement and promote these principles is assumed, which is why they become an integral part of the strategy and organizational culture. Universities seek to promote the principles in their daily work, the leading forums considering the integration in the decision-making process of the principles of sustainability and practices that target the four major areas of interest: human rights, labor, environment and the fight against corruption [1].

Moreover, Global Compact principles and responsible practices are also promoted among stakeholders (partners, students, clients, etc.). Given a large number of stakeholders specific to an academic organization, the training effects nationally and even internationally are considerable. Universities, like other companies that have adhered to these principles, must include in the annual report a description of how these principles have been implemented and how development objectives are supported (Communication of Progress). In this way, the stakeholders are informed and they can make a comparison between the results obtained by the different entities that have adhered to these principles.

Considering the notable differences that exist between different types of organizations as well as the activities carried out in terms of promoting the principles of sustainable development, different reporting standards have emerged internationally. The best known are GRI (Global Reporting Initiative) standards that are used by more than 10,000 entities in over 100 countries.

The GRI company was founded in 1997 after the vehement reactions that appeared worldwide as a result of the environmental damage of the Exxon Valdez oil spill. GRI stood out by launching Guidelines for reporting in 2001 which were periodically improved, with constant concerns eventually leading to the release of GRI reporting standards in 2016. Given the complexity of the phenomenon of promoting sustainable development, GRI has also launched reporting standards on tax transparency. With the help of this standard created in 2019 and which is applicable starting with 2021, entities have the opportunity to properly report relevant information on tax practices to stakeholders, given the importance of financial resources in achieving sustainable development goals [3]. In order to support companies and other entities, GRI launched 2020 a reporting standard on waste impacts based on previous GRI disclosures on waste. Therefore, GRI standards have gradually developed based on international consultations with different categories of stakeholders and their adoption by companies and other entities ensures the possibility of making comparisons on their involvement in promoting the principles of sustainable development.

Many universities use GRI guidelines and standards to prepare sustainability reports. The GRI database query revealed the existence of a number of universities 130 that published 380 reports in the period 1999–2017 using specific guidelines and standards [4], European entities being more concerned by the proper disclosure of information..

Given the specifics of the activities carried out, academic organizations have created various entities as forums for consultation and promotion of sustainable development, on national or international levels [5] like

1. Sustainable Development Solutions Network supported by the United Nations (UN),
2. International Sustainable Campus Network (ISCN),
3. Association for the Advancement of Sustainability in Higher Education (AASHE)
4. Environmental Association for Universities and Colleges (EAUC).

Therefore, universities have at their disposal numerous tools and mechanisms to promote the principles of sustainable development, they stand out not only in the process of use but also in their creation and improvement as they are applied.

This chapter aims to identify the main directions of action of universities in the process of promoting sustainable development. These directions are correlated with the functions that universities fulfil, taking into account the complexity of the sustainable development phenomenon and the different categories of stakeholders involved.

2.2 The Academic Revolutions and Multiple Functions of Universities

The growth of connections in the world economy, the challenges generated by globalization but also climate change have generated the repositioning of universities nationally and internationally. Their activity has become increasingly complex, the last centuries generating significant revolutions in the academic field [2, 6, 7].

The growth of the population, the intensification of the urbanization process, the increase of the complexity of the economic activities have generated the 'massification' of the educational process, which brings new challenges such as covering the funding needs of universities.

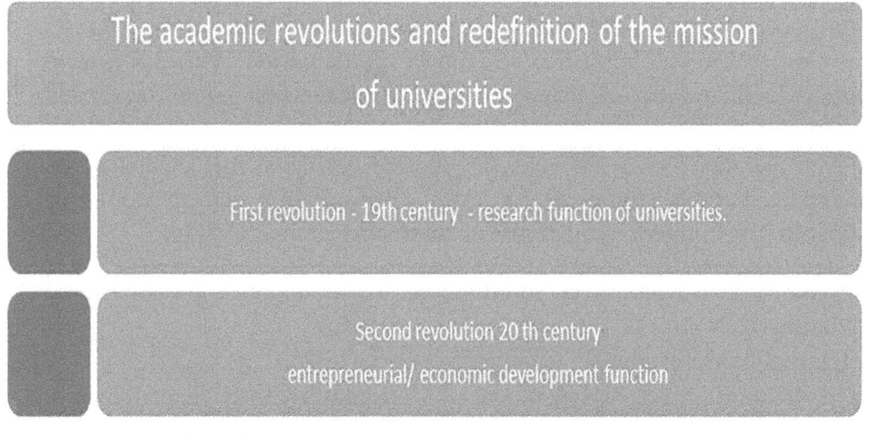

Figure 2.1 Mission of universities [6] Etzkowitz, 2003.

In addition to teaching, research has become increasingly important. The importance of these functions within universities is different, there are teaching universities and research universities, the impact of their activity being thus different in social, economic and scientific terms [8, 9].

The university-company relations have amplified and become more and more complex, the reason for which the universities have become entities that carry out their activity on economic principles, their entrepreneurial spirit involving the companies on a regional and even national level. More and more universities have an entrepreneurial approach to their activity by integrating the function of economic development within the traditional functions of higher education institutions (teaching and research). In this way, universities become leaders of economic development, regionally, nationally and even internationally, by transforming knowledge into capital [2, 6, 10, 11].

Partnerships concluded by universities with various entities such as research institutes, other educational institutions, municipalities, regional authorities, professional associations, chambers of commerce, private companies, banks, NGOs generate the takeover of specific functions to partners and the emergence of new entities. such as incubators, science parks, venture capital funds.

Globalization has also left its mark on the activity of universities that are in a complex process of internationalization. The liberalization of international movements, the international migration, the consecration of English as the dominant language on a scientific level have fuelled the internationalization process of universities which has multiple forms of manifestation [12–14]

In some countries like Qatar, Singapore and the United Arab Emirates, the public authorities have reshaped educational strategies by attracting prestigious foreign universities that have the role of higher education 'hubs' by setting up local campuses [12]. In this way, students have access to a quality educational process, to Western standards and local professors can be attracted to the teaching and research of these universities. In addition, the existence of these foreign universities in the national economic environment generates other positive externalities such as partnerships with local public companies or car companies. In other countries, the instruments of internationalization are different, the universities pursuing the launch and development of study programs in foreign languages and attracting teachers from abroad for both teaching and research. In Europe, the Bologna Process and Lisbon Strategy have a contribution to the establlishment of the European Higher Education Area.

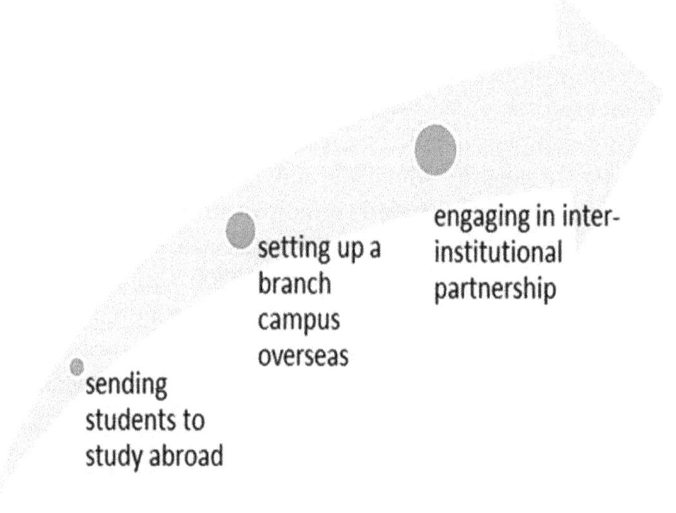

Figure 2.2 Steps for university internationalisation [12].

The internationalization process of the academic world is fed primarily by Information and communications technologies that provide online meetings and data transfer. The COVID crisis has led to the widespread use of new technologies and the creation of virtual research networks that bring together specialists from different countries.

The increasingly complex functions that companies perform, however, are developing under the sign of sustainable development, universities being one of the promoters of this concept through specific channels. Through the didactic function, universities ensure the transmission of the concepts of sustainable development, green economy, corporate social responsibility primarily to students who are future specialists and managers who will ensure the implementation of specific principles at the level of companies or public institutions. The continuous adaptation of the curricula and of the discipline programs ensures the development of specific competencies for the students from the economic faculties, mainly, but also from the technical faculties.

2.3 Moving the SDG 2030 Agenda Forward Through Higher Education Institutions (HEIs)

In 2000, 189 UN member states agreed on the achievement of the Millennium Development Goals (MDGs), 8 specific goals, by the year 2015.

The European Council set a strategic goal for the European Union's spring summit in Lisbon in 2000 – to become the most competitive and dynamic knowledge-based economy in the world that can achieve sustainable economic growth to create more and better jobs and to achieve greater social cohesion [15].

There were many reports and studies done with regard to the performance of each country and monitoring the progress of indicators to achieve these goals [16–21]. The MDGs have been superseded by the Sustainable Development Goals (SDGs), 17 integrated and far more ambitious goals [22].

As shown in Figure 2.3, all these goals are linked with each other and they represent a coherent and meaningful chain of goals related to economy, society and environment. In the centre of the three main pillars of sustainable development goals, it is targeting the goal 17 Partnership, a goal in itself and a goal that reinforces these three pillars. A partnership can be understood as a bridge between universities (networking in the putting together efforts on sustainable development),the collaboration between universities and other actors such as government, businesses and civil societies. Aline with the partnership goal, Boron et al., (2017, p 38) states that the ultimate purpose of academic teaching programs linked and orientated towards sustainability should be the support of practical attainment of a sustainable future for the industry, business and society [23].

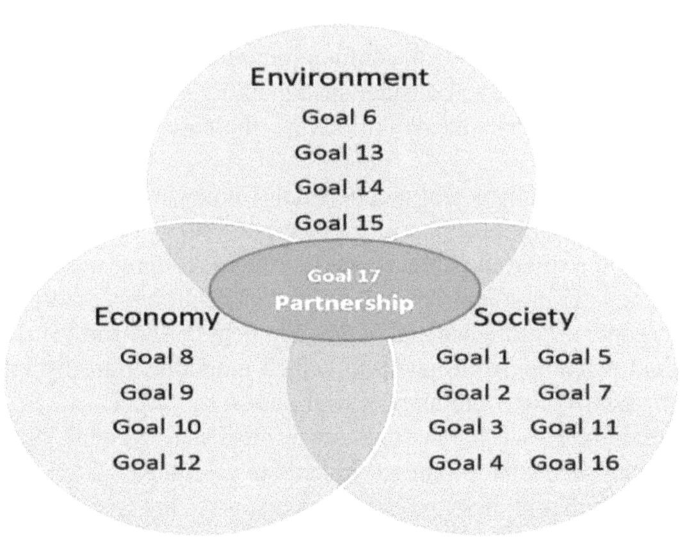

Figure 2.3 Integrated Sustainable Development Goals: Adopted by authors [29].

Literature refers to the triple helix, quadruple helix or even the quintuple helix when discussing collaborations among actors. For instance, according to Etkowitz and Leydersdorff (1995), the quadruple helix is an extension of the triple helix spiral collaboration model of innovation between university–industry–government, which were referred to by them as a 'laboratory for knowledge-based economic development' [24]. Additionally, to achieve open innovation micro- and macro-dynamics, a quadruple helix model is needed for social, environmental, economic, cultural, policy and knowledge sustainability [25].

Higher education for sustainable development is being significantly shaped by the global sustainability agenda [2, 26]. However, we don't always find the fundamentals of sustainable development notions in the higher education system. According to [27] Barth and Reickmann (2012) there is a need for major transformations in higher education systems that would affect all disciplines and levels of study [27]. Furthermore, Franco et al. (2019) states that higher education for sustainable development is a tool addr,essing and supporting the global problems of collaboration and interconnection [26, 28–30]

Generally, as confirmed from missions, visions, and strategic plans from HEIs, they are committed to the societal responsibility of providing guidance to students, faculty, staff and administration and their decisions and 'output' affect economic-social-environmental dimensions of their influencing zones around them [31]. Therefore, we see a direct link to the support in the promotion of sustainability in communities and regions around them [32]. However, if HEIs have to consider sustainability a responsibility rather than a tool for profitability, this is not always the case for some business schools [33].

Moon et al., (2018) developed a conceptual framework that links inputs from higher education to sustainability outcomes [34]. According to their study and based on a rigorous literature review, there are three crucial conditions that pedagogical interventions can impact: (1) entrepreneurship education (ED), (2) education for sustainable development (ESD) and (3) passage from traditional pedagogy to a new one, which embodies transdisciplinary learning, partnerships and economical-social mindsets.

Still, the literature review identifies some mismatch between the goals of traditional entrepreneurship and sustainable development. Often, we see that economic growth and the exploitation of resources are found to be self-interest in the case of entrepreneurship, but limiting growth and conserving resources in the case of targeting sustainability [34–37].

It should be mention that there are other ways of engaging with SDG apart from the integration of concepts into the curricula. For example, signing on international charters or declarations or going through green campus activities are methods of supporting sustainable development [38, 39].

Franco et al., (2019) in their study comprise a significant curriculum mapping throughout four geographical areas (Asia and the Pacific, Americas, Africa and Europe). Findings show that Quality Education, Affordable and Clean Energy, Sustainable Cities and Communities, Responsible Consumption and Production and Life on Land, are covered by the four areas investigated [26]. From these regions, America is showing proactive inclusion of sustainable development thematic in its policy, curriculum and practice. Whereas Africa seems to be less evidence and lack a governing approach to higher education including the sustainable development goals.

2.4 Responsible Management Education and Sustainable Development

Given the complexity of the phenomenon of sustainable development, the existence of trained stakeholders is essential. For this reason, the role of universities is essential in the process of training specialists and managers, conducting scientific research and establishing a dialogue with various stakeholders such as companies, public institutions to improve the social and environmental performance of various entities. The training of specialists must focus not only on the protection of the environment but also on respect for human rights, the fight against corruption, given the devastating effects on stakeholders of the financial scandals in which large transnational corporations have been involved [1, 40].

Managers and business specialists need to become familiar with new and complex concepts such as corporate social responsibility, corporate governance, circular economy, low carbon economy or green finance and which implies a multidisciplinary approach to the educational process [41]. The paradigm shifts that are taking place in the world economy are accompanied by the emergence of new concepts, tools and financial operations that try to facilitate the transition to the green economy.

An essential role in ensuring a general framework for promoting the concept of sustainable development is played by the United Nations, which initially launched the Global Compact principles for companies. Considering the success of these principles and their embrace by other entities, Principles for Responsible Management Education were launched. Currently, over 16,000 business and management programs are part of this network.

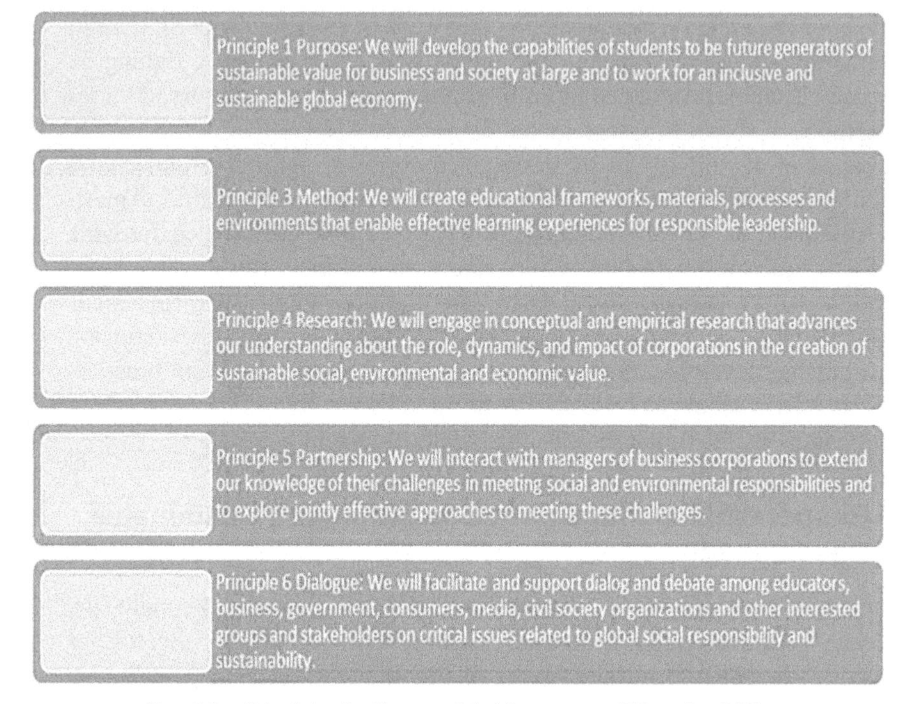

Box 2.1 Principles for Responsible Management Education [42].

Universities that adhere to these principles prepare Sharing Information on Progress (SIP) report in order to transmit relevant information to different categories of stakeholders such as students, employers, graduates, media, etc. These reports present the progress made by universities in promoting the SDGs and the concerted actions undertaken. In addition, they have the role of sharing the experience gained in this complex process with other entities with concerns in this field.

These principles together with the principles of the Global Compact are complex tools through which humanity tries to find solutions to the challenges generated by globalization, global crises, the intensification of industrialization and urbanization processes. The pursuit of profit has generated many negative consequences that are felt economically, socially and environmentally. For this reason, efforts to restore balance are multiple and one of the solutions is education for sustainable development that begins in the family from preschool but is continued throughout life. Universities are the main providers of education for sustainable development both through the bachelor's and master's programs they provide but also through modules focusing on various topics that can be accessed by graduates throughout life. The importance of this type of education is also revealed by the international initiative taken by the Johannesburg World Summit on Sustainable Development (2002). The result was the set up of period 2005–2014 as UN Decade for Education for Sustainable Development. The next step was the UN's Global Action Programme on Education for Sustainable Development (2015–2019). The most important step in recognizing the importance of education for sustainable development is the launch of Sustainable Development Goals (SDG) that are built around education.

2.4.1 Sustainable Development Through Curricula: Some Examples

The topic of sustainable development can be found as a specific course or some related issues can be covered within the curriculum. The first example we consider is a specific course for economics program, 'Development and Growth'. In this course, we see that many aspects of the sustainable development goals are discussed and analysed. Specifically, considering three dimensions of SDGs, economic, social and environmental, Table 2.1 lists the tracks through the topics covered in this course. This table includes information for two specific academic years, that of 2014–2015 and 2019–2020 and the topics are covered. As we see, most of the topics for the three dimensions are significantly considered.

Table 2.1 SDGs though the course 'Development and Growth'

Economic Dimension	Social Dimension	Environmental Dimension
• Continuous improvement in economic well-being (2014, 2019/2020) • Creation of new markets and opportunities for sales growth (2014, 2019/2020) • Cost reduction through efficiency improvev ments (2014) • Greater economic equality (2014) • Efficient use of renewable resources • Protection of commercial rights (2014) • Green business opportunities (2014) • Sharing of commercial information (2014) • Opportunities for reduced energy and raw material input • Fair and equal access to information and knowledge (2014, 2019/2020)	• Protection of human rights • Protection of human health (2014, 2019/2020) • Child mortality • Food security (2014, 2019/2020) • Sharing of vital information (2014, 2019/2020) • Respecting the viewpoints of beneficiaries and victims of developmental activities • Accountability and responsibility of societal decision makers (2019/2020) • Participatory approach in decision making (2014, 2019/2020) • Increasing literacy levels and employment opportunities (2014, 2019/2020)	• Renewal of energy (2019/2020) • Renewal of materials (2019/2020) • Renewable food resources (2019/2020) • Reduction of environmental footprint (2019/2020) • Accountability and responsibility of environo mental decision makers (2014, 2019/2020) • Environmental ethics (2014, 2019/2020)

Source: [43, 44]

Table 2.2 Dimensions of sustainable development and students' project topics

Sustainable Development Dimensions	Project Topics in 2014	Project Topics in 2019/2020
Economic Dimension	1 - Economic Empowerment of Women as Farmers; 2 - Dhurata.com Website, Innovative and Business Development; 3 - Improving the process of collecting milk from farmers; 4 - Made4 Impact.	1 - Entrepreneurial Roadmap Workshop; 2 - Raising awareness of economic issues at a students' club; 3 - Promoting the Tourism Sector of Albania; 4 - Agricultural solidarity and coordination group.
Social Dimension	1 - Eradicate extreme poverty and hunger; 2 - Eat Well, Move More, Live longer; 3 - Alumni Association Formation: A guide for Epoka University.	1 - Gender Equality in Albania: An approach to social, cultural, and financial matters; 2 - Assisting toddlers' development; learning through play; 3 - A "Sharing mean Caring" Journey Contribution to Orphan's Lives; 4 - To care for those who once cared about us, is one of the highest honors; 5 - Bringing Joy to Orphans; 6 - Helping school children with IT skills and online courses.
Environmental Dimension	1 - Ensure Environmental Sustainability by Painting Trees with the Motto "All of us, for a Cleaner Environment".	1 - Water Pollution Awareness Project in Devoll Valley; 2 - "Reduce, Reuse, Recycle" Saving our earth.

Source: [43, 44]

Whereas Table 2.2 gives detailed information on the applied projects of students. They were asked to undergo their concrete contribution to sustainable development. They have selected the topic and specified the target groups that would be affected by their activities and project outcomes.

The integration of applied projects with the taught topics is found to be a good interactive way of engagement in real output to society and to sustainable development.

2.4.2 Sustainable Development Through Green Campus Activities: Some Examples

Bakar et al., (2019) propose the 5S quality management framework for the universities ecosystem to address the challenges with sustainable development, as well as to achieve green campus [45]. The structure of the 5S quality management framework is based on 5 notions: sort, set in order, shine, standardize and sustain. The results reveal the proposed framework would contribute to (1) Reduced waste in all procedures; (2) Increase employee motivation; (3) Increase job quality; (4) Cultivate safety and healthy environment; (5) Improve work culture; (6) Providing positive image to customers.

All the above-mentioned targets are important components of sustainable development that can be the target to higher education institutions.

The study of Tiyarattanachai and Hollmann (2016) compared the perception of stakeholders in Green Campus and Non-Green Campus universities in Thailand with regard to stakeholders 'satisfaction on sustainability practices and at the same time, the perceived quality of life at their campuses [46]. As expected, the results show the universities that adhered to the green campus had a positive effect on sustainability practices and their quality of life. The study also suggests that universities could undergo the criteria set of UI GreenMetric World University Ranking, developed in 2010.

Choi et al., (2017) in their study presented a comprehensive framework of green campus strategies of Portland State University, a higher education institution in Korea [47]. This study included the students' perceptions and their practices in line with the green campus. PSU's sustainable campus plan has been nationally and internationally recognized. The main indicators and investigating issues were the following: plans and goals for green campus; sustainable educational program; sustainable practices and knowledge about green building; features or facilities of the institution green campus; eco-roofs of Cramer Hall and Broadway Housing. Additionally, the student's perceptions and their active inclusion in green campuses, environments and study programs are found sweeping and in-line with the strategies of the institution. At the same time, Portland State University's sustainable campus plan has been nationally and internationally recognized [47] (Choi et al., 2017).

2.4.3 Sustainable Development Through Other Initiatives of Universities

Another form of involvement of universities in promoting sustainable development is the establishment of sustainability centres, through which these entities contribute to the transition to the green economy. Thus, sustainability has been integrated into the mission and strategy of universities and has become the watchword in education, research and innovation. Some researchers even speak of sustainability as the fourth mission of universities [5, 48]. Moreover, given the involvement of universities in local communities and the national economy, sustainable development is 'a social learning process within and beyond academia ' [5, p. 1423].

The study of Sevdalina and Elitsa (2011) examined the relationship between sustainable development and national and international security as an expression of unity between security policy, economy, social environment and ecology [49].

In addition, the universities can also promote the principles of sustainable development as portfolio investors. The big universities have considerable financial resources, which is why they are participants in the capital market, choosing to place their available funds in securities traded on the stock exchange [1, 50–52]. In this way, they act as socially responsible investors who develop their investment strategy based on negative screening which consists in excluding securities issued by companies in certain fields (gambling, tobacco, weapons or nuclear weapons) or countries like Sudan or South Africa.

As the behavior of socially responsible investors has become more sophisticated, universities have also adopted the same trend and academic organizations divest or no longer invest in fossil fuel companies or consider purchasing securities issued by companies that comply with positive ESG (environment, social, governance) standards. Moreover, among the signatories of the United Nations Principles on Responsible Investors are universities such as RMIT University from Australia, University of Edimburg, University of Cambridge.

Investing in the community involves providing capital for communities that are not served to the optimal standards of the financial markets. For example, universities can be involved in funding projects for the community given that they are important sources of cultural and intellectual development. Thus, Ohio State University launched in 1995, an investment initiative for the community for the urban development of the area around the university campus. These investments were financed both from own funds and from funds attracted through the issuance of bonds [53, 54]. In this way, the university's reputation grows and may attract additional donations in the future.

2.5 Conclusion

Universities have become important economic agents in the economy as they expand the functions they perform. The development of society has led to the metomorphosis of universities that have gone through several revolutions that have led to the emergence of functions such as research, entrepreneurship and more recently the promotion of sustainable development.

The involvement of universities in promoting the principles of sustainable development has been achieved gradually, these entities embracing the concept of social responsibility. Social responsibility can be seen as a form of integrated self-regulation of university strategy that seeks to enhance the positive effects that these entities have on the environment, employees, students and the community at large or as a deliberate pursuit of the public interest by academic organizations. The principles of sustainable development are promoted by universities both in their teaching and research activity, as well as in their capacity as entrepreneurial entities. Their involvement in socially responsible investments or impact investments is evidence of this commitment.

The relationship between education and sustainable development is essential, with international forums recognizing the importance of education for the success of promoting the principles of sustainable development at the international level. This is demonstrated by the inclusion of education as a stand-alone goal (SDG 4) on the 2030 Agenda for Sustainable Development and setting up of targets on education under several other SDGs, like those on health; climate change, growth and employment; sustainable consumption and production.

National education systems need to be rethought and reshaped given the challenges facing nations on various economic, social, technical levels. Thus, the education system must prepare students and future employees so that they have the necessary knowledge and skills given the problems facing humanity - climate change, pollution, natural disasters, urbanization, the persistence of poverty in certain areas, health or financial crises. In addition, given the complexity of economic and technical phenomena, it is necessary to learn throughout life, this being a goal of the 2030 agenda with equity and inclusion, gender equality.

Not only companies and public authorities have become aware of their social responsibility, but also universities which, as portfolio investors, can direct their funds not only according to economic criteria but also to considerations related to the environment, society and corporate governance. Thus, a new segment of the capital market has emerged, namely the socially responsible investment market in which large academic organizations in developed countries are important players.

References

[1] Matei, M. (2013). Responsabilitatea socială a corporaţiilor şi instituţiilor şi dezvoltarea durabilă a României. *Bucharest: Expert Publishing House.*

[2] Panait, M., Petrescu, M. G., Podasca, R (2015) Remodeling the role of universities in the context of sustainable development, The International Conference "Economic Scientific research – Theoretical, Empirical and Practical Approaches" – ESPERA 3-4 December 2015, Bucharest, published în *Economic Dynamics and Sustainable Development – Resources, Factors, Structures and Policies* Proceedings ESPERA 2015 – Part 1, p. 375–383

[3] Global Reporting Initiative, 2021, Topic Standard Project for Tax (globalreporting.org)

[4] GRI Sustainability Disclosure Database (2021), SDD - GRI Database (globalreporting.org)

[5] Soini, K., Jurgilevich, A., Pietikäinen, J., and Korhonen-Kurki, K. (2018). Universities responding to the call for sustainability: A typology of sustainability centres. *Journal of Cleaner Production*, vol. 170, pp. 1423–1432.

[6] Etzkowitz, H. (2003). Research groups as 'quasi-firms': the invention of the entrepreneurial university. *Research Policy*, vol. 32, no. 1, pp. 109–121.

[7] Vasile, V., Gh, Z., Perţ, S., and Zarojanu, F. (2007). Restructurarea sistemului de educaţie din România din perspectiva evoluţiilor pe piaţa internă şi impactul asupra progresului cercetării. *Institutul European dinRomânia, Bucureşti.*

[8] Altbach, P. G. (2008). The complex roles of universities in the period of globalization. untitled (upc.edu)

[9] Yasmin, F., Akbar, A., and Hussain, B. (2016). The impact of perceptual learning styles on academic performance of masters'level education students. *Science International*, vol. 28, no. 3.

[10] Zaman, G., and Georgescu, G. (2013). Relaţia cercetare-dezvoltare-inovare şi impactul asupra competitivităţii economice în România în contextul globalizării şi integrării europene, MPRA_paper_52944.pdf (uni-muenchen.de)

[11] Pipirigeanu, M., Zaman, G., Strasser, H., Aramă, R., and Strasser, C. (2014). Academic entrepreneurship and scientific innovation in context of Bio-economy strategy. *Procedia Economics and Finance*, vol. 8, pp. 556–562.

[12] Altbach, P. G., Reisberg, L., and Rumbley, L. E. (2009). Trends in global higher education: Tracking an academic revolution. A Report Prepared for the UNESCO 2009 World Conference on Higher Education

[13] Chen, S., Lin, Y., Zhu, X., and Akbar, A. (2019). Can international students in China affect Chinese OFDI—Empirical analysis based on provincial panel Data. *Economies*, vol. 7, no. 3, pp. 87.
[14] Lin Y, Yasmin F, Chen S. An empirical investigation of the nexus between international students in China and foreign direct investment based on panel threshold model. 2nd International symposium on economic development and management innovation (EDMI 2020), Hothot, China, June 20–21, 2020.
[15] Stoyanova, P. E. (2011). Infrastructures and processes-Europe's supranational cultures. *Review of General Management*, vol. 14, no. 2, pp. 97–104.
[16] Afful-Dadzie, E., Afful-Dadzie, A., and Oplatková, Z. K. (2014). Measuring progress of the millennium development goals: a fuzzy comprehensive evaluation approach. *Applied Artificial Intelligence*, vol. 28, no. 1, pp. 1–15.
[17] Hoxhaj, J., Bllaci, D., Hodo, M., Pici, E., and Hysa, E. (2014). Millennium development goals, MDG'S; Case of Kosovo. *Mediterranean Journal of Social Sciences*, vol. 5, no. 14, pp. 123–123.
[18] Pici, E., Pasmaciu, J., Hysa, E., Hoxhaj, J., and Hodo, M. (2014). Evaluation of millennium development goals process: Case of Albania. *Mediterranean Journal of Social Sciences*, vol. 5, no. 14, pp. 33–33.
[19] Voica, C., Panait, M., Rădulescu, I. (2015) *From Millennium Development Goals to Sustainable Development Goals – Results and Future Developments,* The International Conference "Economic Scientific research – Theoretical, Empirical and Practical Approaches" – ESPERA 3-4 December 2015, Bucureşti, România published in *Economic Dynamics and Sustainable Development – Resources, Factors, Structures and Policies* Proceedings ESPERA 2015 – PeterLang, Berlin, Part 1, pp. 21–32
[20] Durokifa, A. A., and Abdul-Wasi, B. M. (2016). Evaluating Nigeria's Achievement of the Millennium Development Goals (MDGs): determinants, deliverable and shortfalls. *Africa's Public Service Delivery and Performance Review*, vol. 4, no. 4, pp. 656–683.
[21] Brault, M. A., Mwinga, K., Kipp, A. M., Kennedy, S. B., Maimbolwa, M., Moyo, P., ... and Vermund, S. H. (2020). Measuring child survival for the Millennium Development Goals in Africa: what have we learned and what more is needed to evaluate the Sustainable Development Goals?. *Global Health Action*, vol. 13, no. 1, pp. 1732668.

[22] World Health Organization. (n.d.). Retrieved Apr. 20, 2021, from www.who.int/topics/millennium_development_goals/en/.

[23] Boron S, Murray KR, Thomson GB (2017) Sustainability education: towards total sustainability management teaching. In: Filho WL, Brandli L, Castro P, Newman J (eds), Handbook of theory and practice of sustainable development in higher education, vol 1, pp 37–52. https://doi.org/10.1007/978-3-319-47868-5

[24] Etzkowitz, H., and Leydesdorff, L. (1995). The Triple Helix–University-industry-government relations: A laboratory for knowledge based economic development. *EASST review*, vol. 14, no. 1, pp. 14–19.

[25] Yun, J. J., and Liu, Z. (2019). Micro-and macro-dynamics of open innovation with a quadruple-helix model, *Sustainability* 2019, 11, 3301

[26] Franco, I., Saito, O., Vaughter, P., Whereat, J., Kanie, N., and Takemoto, K. (2019). Higher education for sustainable development: actioning the global goals in policy, curriculum and practice. *Sustainability Science*, vol. 14, no. 6, pp. 1621–1642.

[27] Barth, M., and Rieckmann, M. (2012). Academic staff development as a catalyst for curriculum change towards education for sustainable development: an output perspective. *Journal of Cleaner Production*, vol. 26, pp. 28–36.

[28] O'Byrne D, Dripps W, Nicholas KA (2015) Teaching and learning sustainability: an assessment of the curriculum content and structure of sustainability degree programs in higher education. *Sustainability Science* vol. 10, no. 1, pp. 43–59

[29] United Nations (2015) Transforming our World: the 2030 agenda for sustainable development. United Nations, New York

[30] Yonehara A, Saito O, Hayashi K, Nagao M, Yanagisawa R, Matsuyama K (2017) The role of evaluation in achieving the SDGs. Sustain Sci (Spec Feature Sustain Sci Implement Sustain Dev Goals) vol. 12, pp. 969–973

[31] Katiliūtė, E., Daunorienė, A., and Katkutė, J. (2014). Communicating the sustainability issues in higher education institutions World Wide Webs. *Procedia-Social and Behavioral Sciences*, vol. 156, pp. 106–110.

[32] Karatzoglou, B. (2013). An in-depth literature review of the evolving roles and contributions of universities to education for sustainable development. *Journal of Cleaner Production*, vol. 49, pp. 44–53.

[33] James, C. D., and Schmitz, C. L. (2011). Transforming sustainability education: Ethics, leadership, community engagement, and social entrepreneurship. *International Journal of Business and Social Science*, vol. 2, no. 5, pp. 1–7.

[34] Moon, C. J., Walmsley, A., and Apostolopoulos, N. (2018). Governance implications of the UN higher education sustainability initiative. *Corporate Governance: The International Journal of Business in Society.*

[35] Snelson-Powell, A., Grosvold, J., and Millington, A. (2016). Business school legitimacy and the challenge of sustainability: A fuzzy set analysis of institutional decoupling. *Academy of Management Learning & Education*, vol. 15, no. 4, pp. 703–723.

[36] Aragon-Correa, J. A., Marcus, A. A., Rivera, J. E., and Kenworthy, A. L. (2017). Sustainability management teaching resources and the challenge of balancing planet, people, and profits. *Academy of Management Learning & Education*, vol. 16, no. 3, pp. 469–483.

[37] Moon., C.J., (2017), 100 Global Innovative Sustainability Projects: Evaluation and Implications for Entrepreneurship Education, European Conference on Innovation & Entrepreneurship 2017. www.researchgate.net/profile/Christopher_J_Moon/contributions

[38] Sammalisto, K., and Lindhqvist, T. (2008). Integration of sustainability in higher education: A study with international perspectives. *Innovative Higher Education*, vol. 32, no. 4, pp. 221–233.

[39] Lozano, R., and Watson, M. K. (2013). Assessing sustainability in University curricula: Case studies from the University of Leeds and the Georgia Institute of Technology. In *Sustainability Assessment Tools in Higher Education Institutions* (pp. 359–373). Springer, Cham.

[40] Godemann, J., Haertle, J., Herzig, C., and Moon, J. (2014). United Nations supported principles for responsible management education: purpose, progress and prospects. *Journal of Cleaner Production*, vol. 62, pp. 16–23.

[41] Storey, M., Killian, S., and O'Regan, P. (2017). Responsible management education: Mapping the field in the context of the SDGs. *The International Journal of Management Education*, vol. 15, no. 2, pp. 93–103.

[42] www.unprme.org/

[43] Hysa, E. (2014). Defining a 21st century education: Case study of development and growth course. *Mediterranean Journal of Social Sciences*, vol. 5, no. 2, pp. 41–41.

[44] Hysa, E., and Mansi, E. (2020). Integrating teaching and learning in graduate studies: economic development course. *Technology Transfer: Innovative Solutions in Social Sciences and Humanities*, vol. 3, pp. 61–64.

[45] Bakar, N. A., Rosbi, S., Bakar, A. A., Arshad, N. C., Abd Aziz, N., and Uzaki, K. (2019). Framework of 5S Quality Management for University Ecosystem to Achieve Green Campus. *International Journal of Scientific Research and Management*, vol. 7, no. 12.

[46] Tiyarattanachai, R., and Hollmann, N. M. (2016). Green Campus initiative and its impacts on quality of life of stakeholders in Green and Non-Green Campus universities. *SpringerPlus*, vol. 5, no. 1, pp. 1–17.

[47] Choi, Y. J., Oh, M., Kang, J., and Lutzenhiser, L. (2017). Plans and living practices for the green campus of Portland State University. *Sustainability*, vol. 9, no. 2, pp. 252.

[48] Trencher, G., Yarime, M., McCormick, K. B., Doll, C. N., and Kraines, S. B. (2014). Beyond the third mission: Exploring the emerging university function of co-creation for sustainability. *Science and Public Policy*, vol. 41, no. 2, pp. 151–179.

[49] Sevdalina, D., and Elitsa, P. (2011). Sustainable development and National security. *Review of General Management*, vol. 13, no. 1, pp. 44–54.

[50] Vivo, L. A., and Franch, M. R. B. (2009). The challenges of socially responsible investment among institutional investors: Exploring the links between corporate pension funds and corporate governance. *Business and Society Review*, vol. 114, no. 1, pp. 31–57.

[51] Smith, R. L., and Smith, J. K. (2014). Universities and colleges as socially responsible investors. *Available at SSRN 2500438.*

[52] Lord, M. (2020). University endowment committees, modern portfolio theory and performance. *Journal of Risk and Financial Management*, vol. 13, no. 9, pp. 198.

[53] Tekula, R., and Jhamb, J. (2015). Universities as intermediaries: Impact investing and social entrepreneurship. *Metropolitan Universities*, vol. 26, no. 1, pp. 35–52.

[54] Dominguez, O. (2010). Values and responsibility: developing a responsible investment policy at the University of British Columbia (Doctoral dissertation, University of British Columbia). https://open.library.ubc.ca/cIRcle/collections/graduateresearch/310/items/1.0102524

3

Sustainability in Portuguese Higher Education Institutions

Ana Caria[1], Cristiana Leal[1,2], Carolina Machado[1,3], Benilde Oliveira[1], and Lídia Oliveira[1]

[1]School of Economics and Management, University of Minho, 4710-057 Braga, Portugal
[2]Centre for Research in Economics and Management (NIPE), University of Minho
[3]Interdisciplinary Centre of Social Sciences (CICS.NOVA.UMinho), University of Minho
E-mail: aalexandra@eeg.uminho.pt; ccerqueira@eeg.uminho.pt; carolina@eeg.uminho.pt; benilde@eeg.uminho.pt; lidiaoliv@eeg.uminho.pt

Abstract

Higher Education Institutions (HEIs) play a key role in sustainable development. This chapter analyses the path of Portuguese HEIs in pursuing more sustainable practices from the first initiatives to the present time. The Portuguese HEIs have been following international trends for sustainability and developing strategies in the pursuit of the Sustainable Development Goals (SDGs) defined by the United Nations (UN). These institutions have been improving their practices and reporting for sustainable development since mid-2000s'. Additionally, the participation of Portuguese HEIs in several sustainability rankings increased as well as its positioning on those rankings. This chapter describes the main international milestones and how they were interpreted and implemented in the Portuguese context. It focuses primarily on the discussion of the participation of the HEIs in attaining SDGs and their commitment in developing actions towards sustainability; the description of their participation and positioning in the main sustainability rankings; and the analysis of the sustainability reporting of four HEIs that are in front positions as far as Portugal is concerned.

Keywords: Sustainable Development, Sustainable Development Goals, Higher Education Institutions, Sustainability Rankings, Sustainability Reporting, Portugal.

3.1 Introduction

The population growth, the predicted to run out of natural resources and the general environmental deterioration, as well as the worldwide growing economic and social asymmetries have emphasized the limitations of the traditional development model and imposed a new development paradigm. Therefore, SD emerged as the leading model for societal development. SD has been defined from various perspectives, but the most often cited definition is the one proposed in 1987 by the Brundtland Commission Report, which defines SD as the 'development that meets the needs of the present without compromising the ability of the future generations to meet their own' ([1], p. 37). SD is a broad concept anchored essentially on three-dimensional distinct but interconnected pillars, the environmental, economic and social [2, 3]. Consequently, proper development decisions are those that meet the needs of society and are environmentally and economically sustainable, economically and socially equitable as well as socially and environmentally acceptable [4].

This development paradigm has been translated both into the MDGs and the SDGs of the UN. The MDGs sought to guide, in the period 2000–2015, global action to address the basic needs of the world's poorest countries [5]. The SDGs are built on the MDGs but encompass a more diverse range of obstacles and challenges to overcome that is not limited to specific countries but to all countries and within them, the most vulnerable groups. The SDGs were launched in 2015, as part of the UN 2030 Agenda, a global call to action (unanimously adopted by the different world leaders) that aspires by 2030, 'to end poverty and hunger everywhere; to combat inequalities within and among countries; to build peaceful, just and inclusive societies; to protect human rights and promote gender equality and the empowerment of women and girls and to ensure the lasting protection of the planet and its natural resources. We resolve also to create conditions for sustainable, inclusive and sustained economic growth, shared prosperity and decent work for all, taking into account different levels of national development and capacities' ([6], §3). The UN 2030 Agenda sets 17 universal, integrated and indivisible goals (the SDGs), 169 indicators and 223 targets that aim to trigger action in five instances: People, Planet, Peace, Prosperity and Partnerships [6].

The SDGs are a challenge whose success requires the involvement and collaboration of the different stakeholders (e.g. governments, private and public companies, third sector, civil society) worldwide. In this regard, educational institutions and particularly the HEIs play a critical role as they function as the driving forces behind the necessary transformations to be implemented in societies. As a source of promotion, creation and dissemination of knowledge, where education, research and development, innovation are watchwords, HEIs are key players for an effective understanding and reach of the SDGs in local, national and global contexts.

Aware of the importance of HEIs in promoting sustainability and the SDGs in particular, this chapter seeks to address the current state of sustainability, with particular emphasis on the issue of sustainability in higher education in Portugal.

Dividing the chapter into six sections, after the introduction, we come across the second section, which seeks to make a brief presentation on HEIs and sustainability, followed by sustainability in HEIs in Portugal, in section three. With a particular focus on HEIs in Portugal, we have sections four and five. More specifically, dealing with Portuguese HEIs in sustainability rankings, section four begins to present STARS, followed by THE Impact Rankings and ending with GreenMetric World Universities rankings. In turn, section 3.5 draws attention to the sustainability report, highlighting the cases of four Portuguese HEIs, namely, Faculty of Engineering of the University of Porto (FEUP), UMinho, UC and ISCTE-UIL. The chapter ends with a brief conclusion highlighting the main ideas worked throughout it.

3.2 Higher Education Institutions and Sustainability

Education are pointed out as one of the most powerful and effective vehicles for SD [6]. As stated by Irina Bokova, Director-General of UNESCO from 2009–2017 ([7], p. 16), 'economic and technological solutions, political regulations or financial incentives are not enough. We need a fundamental change in the way we think and act'. Adding to primary and secondary schools, HEIs have also been called to contribute to the construction of a more inclusive and sustainable society. These institutions have a role to play in the quest for SD and to achieve the SDGs for 2015–2030. Education and scientific research are mentioned in several SDGs (SDG 4- Quality Education; SDG 8- Decent Work and Economic Growth; SDG 10- Reduced Inequalities and SDG 17- Partnerships for the Goals), however, to achieve compliance with the SDGs the involvement of HEIs has to be comprehensive [8, 9].

The involvement of HEIs with sustainability issues started to assume particular prominence with the United Nations Decade of Education for Sustainable Development 2005–2014 (UN DESD) [7]. Nonetheless, since the early 70s, several forums/conferences, diplomas, objectives and measures have been progressively directed towards the education sector, aiming to promote and develop a culture of sustainability. Below some of the main milestones of the DESD are listed:

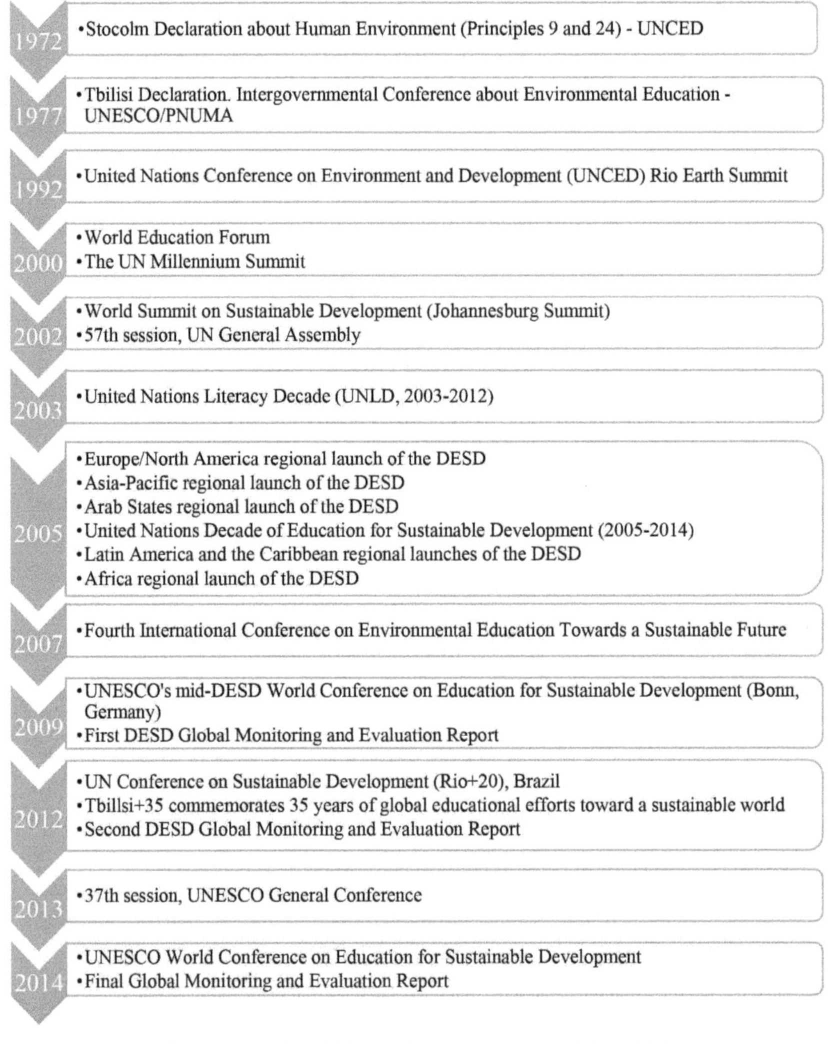

Figure 3.1 The DESD Milestones (Adapted from [7]).

These milestones are important indicators of the efforts that have been made during these times within the scope of the ESD, aiming to stimulate a set of transformations in the field of education from which we seek to contribute to the reorientation of societies towards SD.

HEIs are demanded to integrate sustainability into their visions, goals and practices and to contribute - through teaching and learning, research, operations and governance and, external leadership[1] - to translate sustainability principles to future regenerative societies and to bring global issues closer to students and future leaders [13, 16–18]. HEIs should not only teach content on SD but educate for SD [19, 20]. The activities and expertise of the HEIs are fundamental to address the different and interconnected challenges related to social development, environmental protection and economic growth of the SDGs. As so, according to the SDSN Australia/Pacific [16], the major roles for HEIs are to:

1. provide the knowledge and solutions that support the implementation of the SDGs;
2. create (current and future) implementers of the SDGs;
3. incorporate the principles of the SDG through organizational governance, management and culture; and
4. provide cross-sectoral leadership in implementation.

Considering education as the most powerful weapon in the transmission of knowledge and in the construction of mentalities, HEIs play a critical role as key players in the promotion and dissemination of sustainability dynamics in the social environment in which they are inserted. As Farinha, Azeiteiro and Caeiro ([21], p. 913) argue 'It is a generally accepted view that

[1]The contribution of HEIs to SD is traditionally identified in the literature as coming from four different areas: Learning and Teaching (Education), Research, Campus Operations and Community Outreach (e.g. [10, 11]). However, given the recognition of the importance of reporting sustainability practices [12, 13], academics have been considering additional areas of HEIs activity (e.g. [14, 15]). [14] and [15] propose the addition of the communication and disclosure of SD practices as a fifth activity. This fifth activity involves the HEIs communication with the different stakeholders through education, research, operations on campus, community outreach and raising awareness in the community (e.g. evaluation and SD reporting). The report 'Getting started with the SDGs in Universities' of the Sustainable Development Solutions Network [17] identifies the following areas of contribution of universities to the SDGs: Learning and Teaching; Research; External Leadership; and Organizational Governance, Culture and Operations. This late vector urges universities to incorporate the SDGs into the organizational reporting.

universities have played a key role in transforming societies, by educating decision-makers, leaders and entrepreneurs'.

As stated, education to sustainability implies the comprehensive contribution of teaching and learning activities, research, operations and governance and, external leadership. In what relates to teaching and learning, SD topics and analysis can be included in academic curricula and research projects to help developing student's global awareness and ability to collaborate and lead SD-related actions in their field of expertise [20]. But more than a simple complement to existing curricula, ESD targets the teaching and learning system itself. The enhancement of participative teaching and the adoption of more creative and innovative learning methods (such as encouraging critical thinking, the ability to see different scenarios, boosting teamwork, group decision-making, among others), foster the adoption of practices and actions oriented to the promotion of SD. In this sense, and for effective achievement of the proposed objectives, it is essential an adjust and reorientation teaching systems and educational structures [7].

Research activities are another important way through which HEIs can contribute to the SDGs, as research-related activities will provide the necessary knowledge, evidence-based, applications and innovations to implement SDG solutions [22, 23]. Additionally, HEIs should set the example of a more sustainable life for other sectors and businesses. HEIs can implement SGD principles in the management of campi and operations; the campus planning, design, construction and rehabilitation of buildings and infrastructures; the purchasing practices; mobility and involvement with the community [24, 25]. To successfully achieve SDGs, the HEIs efforts have to be combined with society at large. HEIs should leverage their neutral and trusted position within society to strengthen public engagement and participation (1) in advocating for SD and (2) in addressing the SDGs, to facilitate cross-sectoral dialogues and partnerships [16].

In this sense, HEIs are asked to show their commitment to sustainability by signing charters, declarations and partnerships such as; the Agenda 21 Report of the UN Conference on Environment and Development, Copernicus Charta 2.0, Declaration of Barcelona, Global Higher Education for Sustainability Partnership, Graz Declaration on Committing Universities to SD 2005, G8 University Summit, Halifax Declaration, Lüneburg Declaration on Higher Education for Sustainable Development, Rede Campus Sustentável, Talloires Declaration, The Magna Charta of European Universities, The UN Decade Education for Sustainable Development 2005–2014, UN Higher Education Sustainability Initiative and the World Declaration on Higher Education for the 21st century: Vision and Action [26]. The signing of sustainability charters or declarations is of relevance as this commitment signals the intention to act in a way that is informed by the SDGs and that could be a catalyst for more sustainability actions [13, 24].

To assess HEIs against their pledges to implement SDGs and how their intent will be backed with action several assessment tools for higher education have been developed. These tools that usually take the form of a ranking, must be able to evaluate the implementation of sustainability in HEIs, which implies identifying the central themes; be based on indicators and conceptual models that support sustainability decisions; be measurable and comparable; go beyond eco-efficiency; be able to evaluate progress and motivations in sustainability dimensions; be understandable to a wide audience and be used as benchmarking practices [27–30]. Some of the sustainability assessment and benchmarking tools available are: the THE Impact Rankings; STARS; the GreenMetrics University Ranking; Assessment System for Sustainable Campus; the Alternative University Appraisal; Plan Vert; the People and Planet University League; the Sustainability Assessment Questionnaire; the Sustainability Leadership Scorecard; the Program Sustainable Assessment Tool and the Unit-Based Sustainability Assessment tool [26, 29]. More detail on this issue, particularly on how Portuguese HEIs position themselves in some of these assessment tools is given in section 4 of this chapter.

In addition to the commitment to become sustainable, HEIs also need to be held accountable and report accordingly. Reporting is not only relevant to communicate with internal and external stakeholders but is also an important measure for diagnosis, evaluation, transparency and accountability of the adoption of sustainability culture. Reporting on SGDs is voluntary but similarly to other sectors (particularly the business sector), HEIs have started to issue standalone reports in sustainability or to incorporate reporting on the SGDs into their annual accounts. The report of the SDSN Australia/Pacific ([16], pp. 40–42) identifies SDGs reporting as a subject of active development and states that HEIs can develop an approach to reporting that suits their own values, priorities and current reporting activities, giving the following suggestions for start reporting on SGDs:

1. identify or develop measures or indicators (e.g. on actions taken, initiatives and their impact, resources developed) to demonstrate the institution engagement with the SDGs and report on them annually;
2. build on existing reporting obligations to develop a single overall narrative of the HEI impact on SDGs;

3. report in a substantive and reliable way, avoiding 'SDG-washing' (not only in areas that the HEI performs well);
4. do not report just for the sake of reporting (define clearly the objective of reporting to identify which strategy is best for the institution).

Although there are no mandatory and comprehensive guidelines and tools that HEIs could draw on to report on their SDG impact, HEIs can opt to follow the GRI standards. GRI is an international leading organization that provides guidelines to report on sustainability and is generally considered the most reliable and accepted reporting framework to follow [31]. The GRI guidelines allow for a choice among three reporting levels that differ with respect to the types of issues and the number of parameters - referred to as performance indicators - that must be reported. While not specifically designed for HEIs, these guidelines have been used by these institutions to report on sustainability. Larrán Jorge, Andrades Peña, and Herrera Madueño [32] identify a total of 138 sustainability reports of 58 universities around the world (mostly European public and larger universities) that follow the GRI guidelines. Another framework that could be followed to report sustainability and HEIs' value creation in the short, medium and long term is the one proposed by The IIRC. IIRC proposes a reporting approach under which financial and sustainability information is presented in an integrated manner [33]. Irrespective of the chosen reporting framework, as suggested by Ramísio, Pinto, Gouveia, Costa and Arezes [25], reporting on sustainability signals and deepens the commitment to sustainability.

In this context, given the importance that HEIs assume in this entire context, it is important to know better the role that HEI in Portugal has, in terms of promoting sustainability before and after 2015, the turning point in the transition from the MDGs to the SDGs.

3.3 Sustainability and HEIs in Portugal

Following the principles and objectives established within the scope of the DESD, in June 2005, in Portugal, the UNESCO National Commission constituted a Working Group, composed of representatives of several entities (public administration and civil society), among which higher education, looking to define a set of contributions to boost the UN DESD in Portugal [34]. In this context, among other reflections, this Group started by emphasizing that Portugal was behind concerning SD, not having, at the time, integrated into its civic, economic and political culture, as well as in its dynamics of action, the values underlying SD. At the same time, they also observed that in Portugal, the objectives of a policy for SD still faced important obstacles, namely concerning the resistance of political forces in relation to policies that did not provide economic growth results in the short term. Aware of this reality and to answer to the challenge launched by the UN, to which Portugal had officially subscribed, a set of objectives were established for the UN DESD in Portugal, namely ([34], p. 8):

1. 'To value the fundamental role that education and learning play in the common search for Sustainable Development;
2. Facilitate relationships and networking, exchange and interaction between ESD stakeholders;
3. Provide a space and opportunities to improve and promote the concept of Sustainable Development as well as the transition to that development through all kinds of awareness and learning for citizens;
4. Participate in improving the quality of teaching and learning in the field of Education for Sustainable Development;
5. Develop strategies, at all levels, to strengthen capabilities in Education for Sustainable Development'.

Following these objectives and with regard to changes (strategic and operational) to be implemented, this Working Group defined, among others, the need to [34]:

1. 'encourage multidisciplinary scientific research in the area of Sustainable Development and Education for Sustainable Development;
2. transform the school (in its different levels of education) into a unit of production and dissemination of information on Sustainable Development and Education for Sustainable Development at local and national level, as well as an intervention agent and a motor of mobilization of society through the students, their families and the rest of the educational community' (p.14);
3. 'Create specific skills in the field of scientific dissemination and communication through postgraduate training initiatives, as well as the creation of communication offices in university and research institutions;
4. Equate and integrate topics of Education for Sustainable Development and its inherent values in the curricula at all levels of education;
5. Make the practice of projects within the scope of Education for Sustainable Development a routine, carried out by schools of all levels of education and involving the community' (p.17);

6. 'Formally insert sustainability topics in the curriculum of degrees in social communication, journalism, public relations, advertising and marketing - either through specific courses, modules of these courses or workshops'; (p.18)
7. 'Create a 'Guide to Good Sustainability Practices' to be adopted by schools and universities, which regulates their daily functioning; item Organize a self-assessment exercise by the different degrees on the level of existence of the themes of Sustainable Development and Education for Sustainable Development in their curricula (with special relevance for degrees that lead to teacher training). Discuss, make recommendations and disseminate the results of this exercise' (p.19).

Finally, it was also highlighted by this Working Group that, in Portugal, the most adequate infrastructure to act as a lever for DESD were schools in their different levels of education. Since HEIs are one of the key elements that are part of these infrastructures, it is observed that, despite the strategies, policies and actions outlined (among which those focused on in the previous paragraph), HEIs still show many gaps. An example of this are the results of some of the studies carried out by different researchers, which highlight the delay that, in this decade, still exists in Portuguese HEIs.

To Farinha, Azeiteiro and Caeiro ([21], p. 913) 'education and research on sustainability and inclusive development in universities is still at an early stage in many institutions, even in Europe, in particular, Southern Europe'. According to the authors, this finding is in line with Aleixo et al., [35] which conclude that more than 50 % of Portuguese HEIs are still in early stages of the process, classifying them as 'laggards' and a 'late majority' in implementing SD.

Still, according to Farinha, Azeiteiro, and Caeiro [21] and based on a set of 7 interviews implemented with key actors in decision-making processes at the level of Portuguese ESD, universities did not define strategies and policies related to SD. 'The government, ministry and CRUP did not have an active role and did not produce any integrator document in the sustainability implementation into policies and strategies of universities' ([21], p. 936).

In another study, Farinha, Caeiro and Azeiteiro [36] based on an analysis of 139 documents from fourteen Portuguese universities, concluded that given that the documents analysed were about integration and environmental education, it might seem that universities were not sufficiently involved in SD during the UN DESD 2005–2014. However, in the analysed documents, some references to actions to implement sustainability in public universities were found. At this level, it is important to emphasize that, the results obtained by these researchers showed that the movement advanced at the university level, with positive examples and initiatives in several Portuguese universities (namely in what concerns sustainability interdisciplinary curricula, specially at the post-graduate level), despite the insufficiency of national strategies and policies combined with or related to ESD.

With a particular focus on the environmental issue, Matos, Cabo, Ribeiro and Fernandes [37] emphasize that in Portuguese higher education the sustainability process of HEIs and courses is still residual. 'According to the consulted documentation, even though there are institutions that have embraced this process and incorporated more efficient management models, there are deficiencies in the global and articulated application of sustainability, within the scope of its basic functionalities: teaching, research and extension' ([37], p. 14). According to these researchers, even though the HEIs' discourse is oriented towards what is defined in international agreements, the truth is that it has not been consistent with practice, particularly concerning the inclusion of the principles of Environmental ESD in teaching-learning. This same conclusion was found in Aleixo, Azeiteiro and Leal [38] when they refer that 'Portuguese HEIs are mainly engaged in the social dimension of sustainability. The economic dimension emerges in second place and the institutional in third; the environmental dimension is the least developed' (p. 146).

Although in a limited number, the identified studies allow us to verify that in the UN DESD 2004–2015, in Portugal, the role played by HEIs in relation to SD was shown to be very limited. In fact, although some HEIs develop some initiatives related to SD, coordination and communication at the national level stood, in this decade, below what was intended.

During the 2030 Agenda and, consequently, in the definition of the 17 SDGs and respective goals to be achieved, on October 31, 2019, in the context of the 1st CCS 2019, held in Portugal, hundreds of representatives, that includes 28 rectors/ presidents, from the main Portuguese HEIs, signed a Letter of Intent, entitled 'Commitment of Higher Education Institutions to sustainable development' aiming to implement a 'Culture of sustainability' in the main national HEIs.

Aware of the importance that HEIs in Portugal have in contributing to the evolution of a sustainable society, with respect for nature and the human person, as well as the need to integrate, across their different activities, the principles underlying the 2030 Agenda and, in this way, to reach the SDGs, the signatory HEIs of this Letter of Commitment committed to developing a set of actions, namely [39]:

1. Institutional commitment;
2. Promotion of ethics for sustainability;
3. Offer of training for sustainability;
4. Transdisciplinarity;
5. Dissemination of knowledge;
6. Collaborative networks;
7. Partnerships;
8. Technology transfer

It should be noted that the underlying concerns here had as their embryo the Sustainable Campus Network, created in 2018 at the Sustainable Campus Meeting, at the University of Coimbra, with the presence of different HEIs interested in promoting the operationalization of activities at the level of sustainable planning and management of their campi. The Sustainable Campus Network assumed, at this stage, the need to promote the involvement of the governance structures of the Portuguese HEIs, as these are considered key elements in the promotion and dissemination of sustainability, and consequently in the operationalization of the SDGs, not only on their campi, but also in the surrounding society. In fact, as HEIs are entities that train senior staff, who develop scientific research through the use and management of public resources, their role as agents promoting the SDGs is particularly prominent. At the same time and also to reinforce this involvement, it is observed that some HEIs, at the level of their governing bodies, had already defined sustainability policies and practices, incorporated in their institutional planning and management, while at the same time creating structures/units capable of operationalizing programs and actions to make the previously established plans viable [39].

Farinha, Azeiteiro and Caeiro [21] and Aleixo, Azeiteiro and Leal [40] observed that the Portuguese HEIs, based on their autonomy and social responsibility, have been developing several initiatives and policies within the scope of EDS. In particular, Aleixo, Azeiteiro and Leal [40] state that 'Portugal has recently published a Green Paper on Social Responsibility and Higher Education Institutions (ORSIES - Observatório da Responsabilidade Social e Instituições de Ensino Superior, 2018) with the support of the State Secretariat for Science, Technology and Higher Education. This document places SDGs at the center of HEI practices, namely in the formative offer domain and therefore signals the recent change in Portuguese Public Policies to promote the ESD. Nowadays, there is debate on how SD should be assessed and reported by HEIs.'

3.4 Portuguese HEIs in Sustainability Rankings

Rankings for sustainability in HEIs are recognized standards that allow the measurement of the sustainability performance, setting goals for long-run continuous improvement and track evolution. Rankings can work as a driving force towards sustainability. The HEIs feel increasing pressure to participate and adequately report on what they are doing. It is not enough to have sustainable actions and practices. Sustainability practices must be visible to measure, report and disclose to all stakeholders [27–30]. Participation in rankings is helpful to set long-term sustainability goals and set sustainability priorities. It identifies the HEIs that are leading the process - usually the ones that started earlier - and those that are still taking their first steps. More and more HEIs enter the race each year and having started earlier or being ahead is no guarantee of continuity in the forefront [29, 30].

Several dimensions are under analysis, such as social responsibility, quality of scientific research, academic excellence and sustainability practices that will determine the classifications of HEIs and, therefore, their prestige. With several parallel rankings, each with specific requirements and methodologies, HEIs must decide in which to participate and in which they have the greatest advantage considering their current position. For instance, STARS can be very helpful for those that are taking the first steps toward sustainability. The THE Impact Rankings brings a broad perspective on the contribution of HEIs considering each of the SDGs while the GreenMetric World Universities makes the analysis focused on sustainability issues and its teaching in a more restricted view [30].

This section provides an overview of the positioning of Portuguese HEIs in main rakings that emphasize those that are in the forefront and those that are taking the first steps. It can also be seen as an instrument to guide sustainable strategic planning to improve positioning. One aspect emerges clearly: this is a differentiating factor of institutional prestige and being left out will have high reputational costs.

3.4.1 Sustainability Tracking, Assessment and Rating System

STARS from the Association for the AASHE is a self-reporting system to measure sustainability performance. The system is useful both for

Table 3.1 Portuguese HEIs using STARS

HEI	STARS Version	Rating	Valid Through
Aberta University	2.1	Bronze	April 4, 2022
University of Porto			

Source: AASHE https://reports.aashe.org/institutions/participants-and-reports/?sort=country, consulted in 6-7-2021.

recognizing high-performing institutions and as a starting point towards the sustainability of institutions that are taking their first steps.

'STARS is designed to:

1. Provide a framework for understanding sustainability in all sectors of higher education.
2. Enable meaningful comparisons over time and across institutions using a common set of measurements developed with broad participation from the international campus sustainability community.
3. Create incentives for continual improvement toward sustainability.
4. Facilitate information sharing about higher education sustainability practices and performance.
5. Build a stronger, more diverse campus sustainability community'[2].

Two Portuguese Universities have registered to use the STARS Reporting Tool: the Aberta University and the University of Porto. Aberta University has earned a STARS rating of Bronze, with a score of 43.00. The most valued dimensions for obtaining this score were curriculum (16.77/34.00) and research (15.80/18.00), both in the academic dimension, as well as campus engagement (9.45/21.00) and public engagement (9.50/18.00), in the engagement dimension [40]. The University of Porto does not achieve a STARTS certification. However, its participation as STARS reporter signals the commitment to a sustainability strategy and the improvement of practices.

3.4.2 Times Higher Education Impact Rankings

The THE Impact Rankings scope is much broader and assesses the contribution of HEIs to each SDG. It is based on four broad areas: research, stewardship, outreach and teaching and it is the one in which more Portuguese HEIs are classified (11 HEIs in 2021). Table 3.2 describes the items under analysis to assess the level of compliance to each SDG.

[2]STARS, https://stars.aashe.org/about-stars/, consulted in 6-7-2021.

Table 3.2 Measurement Method of the THE Impact Rankings to assess HEIs

#	SDG	Measurement elements
SDG 1	no poverty	measures universities' research on poverty and their support for poor students and citizens in the local community
SDG 2	zero hunger	measures universities' research on hunger, their teaching on food sustainability, and their commitment to tackle food waste and address hunger on campus and locally
SDG 3	good health and well-being	measures universities' research on key diseases and conditions, their support for healthcare professions and the health of students and staff
SDG 4	quality education	measures universities' contribution to early years and lifelong learning, their pedagogy research and their commitment to inclusive education
SDG 5	gender equality	measures universities' research on the study of gender equality, their policies on gender equality and their commitment to recruiting and promoting women
SDG 6	clean water and sanitation	measures universities' research related to water, their water usage and their commitment to ensuring good water management in the wider community
SDG 7	affordable and clean energy	measures universities' research related to energy, their energy use and policies, and their commitment to promoting energy efficiency in the wider community
SDG 8	decent work and economic growth	measures universities' economics research, their employment practices and the share of students taking work placements
SDG 9	industry, innovation and infrastructure	measures universities' research on industry and innovation, their number of patents and spin-off companies and their research income from industry
SDG 10	reduced inequalities	measures universities' research on social inequalities, their policies on discrimination and their commitment to recruiting staff and students from under-represented groups
SDG 11	sustainable cities and communities	measures universities' research on sustainability, their role as custodians of arts and heritage and their internal approaches to sustainability
SDG 12	responsible consumption and production	measures universities' research on responsible consumption and their approach to the sustainable use of resources
SDG 13	climate action	measures universities' research on climate change, their use of energy and their preparations for dealing with the consequences of climate change.
SDG 14	life below water	measures universities' research on life below water and their education on and support for aquatic ecosystems.

Continued

Table 3.2 Continued

#	SDG	Measurement elements
SDG 15	life on land	measures universities' research on life on land and their education on and support for land ecosystems.
SDG 16	peace, justice and strong institutions	measures universities' research on peace and justice, participation as advisers for government and policies on academic freedom.
SDG 17	partnerships for the goals	looks at the broader ways in which universities support the SDGs through collaboration with other countries, promotion of best practices and publication of data.

Source: THE –Impact Rankings
www.timeshighereducation.com/rankings/impact/2021/overall#!, consulted in 6-7-2021.

Table 3.3 presents the participation, rank and overall scores of the Portuguese HEIs in THE Impact Rankings, with the University of Coimbra top-ranked in the last 2 years. In 2021, there are 11 institutions in the ranking, one more than in the previous year and seven more than in 2019. This evolution in participation brings evidence that the HEIs recognize the relevance and status of standing in this ranking. The improvement in scores also shows the effort of HEIs to continuously improve their contribution to the SDGs.

The overall score reflects the assessment of the institution's contribution to each of the SDGs. The contribution to the various SDGs can be very different from HEI to HEI depending on the characteristics of the university itself but there are SDGs to which HEIs typically contribute more strongly. Therefore, highlighting the SDGs in which HEIs tend to concentrate better scores, as well as those where there is wide dispersion among HEIs, is essential to strategically decide where to focus efforts.

Table 3.4 presents the detail of the SDGs' best score rank for each institution, considering the four best scores per HEI. The SDG 17 - partnerships for the goals - which looks at the broader ways in which HEIs support the SDGs through collaboration with other countries, promotion of best practices and publication of data is the one that consistently tends to generate high scores for every Portuguese HEIs, a tendency that also occurs in HEIs worldwide. Other SDGs in which HEIs typically score the highest are the SDG 3 – good health and well-being – which measures universities' research on key diseases and conditions, their support for healthcare professions and the health of students and staff; SDG 4 – quality education - which measures universities' contribution to early years and lifelong learning, their pedagogy research and their commitment to inclusive education; SDG 9 – industry, innovation

Table 3.3 Portuguese HEIs in the THE Impact Rankings

Year	#	Rank	Name	Overall Scores
2021	1	21	University of Coimbra	92.7
2021	2	53	NOVA University of Lisbon	89.4
2021	3	101–200	University of Algarve	77.5–85.2
2021	4	101–200	University of Aveiro	77.5–85.2
2021	5	101–200	University of Minho	77.5–85.2
2021	6	101–200	University of Trás-os-Montes and Alto Douro	77.5–85.2
2021	7	201–300	ISCTE-University Institute of Lisbon	71.0–77.4
2021	8	301–400	Catholic University of Portugal	66.3–70.9
2021	9	401–600	Universidade Aberta	56.6–66.2
2021	10	401–600	University of Beira Interior	56.6–66.2
2021	11	601–800	Polytechnic Institute of Setúbal	47.6–56.5
2020	1	62	University of Coimbra	86.5
2020	2	101–200	University of Aveiro	75.4–83.3
2020	3	101–200	University of Minho	75.4–83.3
2020	4	101–200	NOVA University of Lisbon	75.4–83.3
2020	5	201–300	University of Algarve	68.2–75.3
2020	6	201–300	University of Beira Interior	68.2–75.3
2020	7	201–300	University of Trás-os-Montes and Alto Douro	68.2–75.3
2020	8	301–400	Catholic University of Portugal	61.5–68.0
2020	9	301–400	ISCTE-University Institute of Lisbon	61.5–68.0
2020	10	601+	Polytechnic Institute of Setúbal	9.5–46.6
2019	1	83	University of Minho	77.8
2019	2	101–200	University of Aveiro	64.6–75.6
2019	3	101–200	NOVA University of Lisbon	64.6–75.6
2019	4	201–300	ISCTE-University Institute of Lisbon	53.7–64.5

Source: THE Impact Rankings
www.timeshighereducation.com/rankings/impact/2021/overall#!, consulted in 6-7-2021.

and infrastructure – which measures universities' research on industry and innovation, their number of patents and spin-off companies and their research income from the industry; and SDG 16 – peace, justice and strong institutions - which measures universities' research on peace and justice, participation as advisers for government and policies on academic freedom[3]. Table 3.4 identifies the best of each HEI. Although HEIs are making efforts in all SDGs, in some cases those efforts are still very incipient.

[3]https://www.timeshighereducation.com/rankings/impact/2021/overall#!/page/0/length/25/sort_by/rank/sort_order/asc/cols/undefined,consultedin6-7-2021.

Table 3.4 Best SDG Scores by Portuguese HEIs in the THE Impact Rankings

Year	HEI	SDGs – the four best scores per HEI																
		1	2	3	4	5	6	7	8	9	10	11	12	13	14	15	16	17
2021	University of Coimbra		■	■						■								■
2021	NOVA University of Lisbon			■						■							■	■
2021	University of Algarve										■					■	■	■
2021	University of Aveiro						■								■	■		■
2021	University of Minho				■				■	■								■
2021	University of Trás-os-Montes and A.D.		■					■								■		■
2021	ISCTE-University Institute of Lisbon									■			■				■	■
2021	Catholic University of Portugal			■							■						■	■
2021	Universidade Aberta					■			■		■							■
2021	University of Beira Interior			■					■				■					■
2021	Polytechnic Institute of Setúbal	■		■	■													■
2020	University of Coimbra	■		■						■								■
2020	University of Aveiro						■			■						■		■
2020	University of Minho				■					■			■					■
2020	NOVA University of Lisbon					■				■							■	■
2020	University of Algarve	■			■	■												■
2020	University of Beira Interior		■	■					■				■					■
2020	University of Trás-os-Montes and A.D.		■	■	■													■
2020	Catholic University of Portugal								■			■					■	■
2020	ISCTE-University Institute of Lisbon									■		■	■					■
2020	Polytechnic Institute of Setúbal			■	■	■												■
2019	University of Minho			■	■							■						■
2019	University of Aveiro								■			■	■					■
2019	NOVA University of Lisbon			■	■												■	■
2019	ISCTE-University Institute of Lisbon				■	■											■	■

Source: THE Impact Rankings
www.timeshighereducation.com/rankings/impact/2021/overall#!, consulted in 6-7-2021.

Graph 3.1 presents the evolution of the average scores in each SDG by year, determined only for the HEIs that are rated in that SDG. As the THE Impact Rankings uses calibrated indicators some HEIs can perform well in some SDG without being assessed in others. The graph shows that scores have improved in almost all SDGs. On the top of every bar column in the graph is presented the number of observations (HEIs) assessed. It is noteworthy to notice that having more HEIs entering the ranking and more HEIs evaluated in that SDG can lead to a drop in the average score, but despite

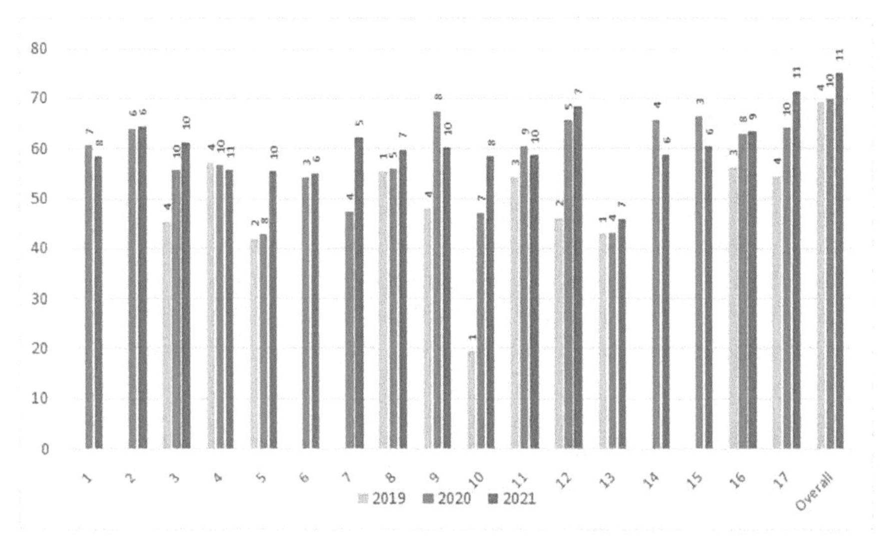

Graph 3.1 Average Score of Portuguese HEIs by SDG and year, in the THE Impact Rankings.

The graph considers the HEIs in the ranking scored in each SDG that year. When the HEI is classified in an interval, the average value was considered. The number of HEIs assessed in that SDG in that year are presented on the top of each bar. SGDs 1, 2, 6, 7, 14 and 15 are not assessed by THE – Higher Education Rankings in 2019.

Source: THE – Higher Education Rankings
www.timeshighereducation.com/rankings/impact/2021/overall#!, consulted in 6-7-2021.

this, the HEIs already in the ranking is consistently improving every year. The average scores evidence that the better scores are obtained in SDG 2, 9, 12, 16 and 17. However, considering the total scores of HEIs in each SDG, instead of their average, the ones with better scores would be SDG 3, 4, 9, 11, 16 and 17. Summing up all HEIs' scores in the 3 years of analysis, these SDG reach cumulative results above 1,000 points - produced either by individual higher scores or more HEIs scoring in that SDG. Anyway, this can also be understood as an effort and contribution to these SDG.

Overall, we observe a significant and continuous improvement toward SD. Both the increase in participating HEIs and the improvement in rankings reveal an interest in contributing as a change agent for SD that allows HEIs to set goals for long-run continuous improvement and develop strategic planning accordingly.

3.4.3 GreenMetric World Universities Rankings

The GreenMetric World Universities Rankings scope is more focused on campus sustainability and assesses three broad areas: environment, economy and equity. The ranking was launched in 2010 and the first Portuguese HEI was included in 2013. Currently, there are four participating HEIs: the University of Minho, the Polytechnic Institute of Viana do Castelo, the University of Aveiro and ISCTE-University Institute of Lisbon. Table 3.5 presents the Portuguese HEI in the GreenMetric Rankings, their position in the world rank, total scores e category score, per year. It shows that the first enters are not currently in the rank, with other HEIs entering and occupying front positions.

Overall, the sustainability competition is soaring. As more and more HEIs are included in the ranking every year, even improving the score - which means getting better - several institutions are surpassed and worsen their position in the ranking. The pressure for sustainability is escalating and even those who are doing well, feel the pressure to do even better if they don't want to be left behind. The UMinho, the University of Aveiro and the ISCTE-UIL are the ones that currently occupy leading positions both in GreenMetric World Universities Rankings and in THE Impact Rankings.

3.5 The Sustainability Report: the Cases of Four Portuguese HEIs

In this section, through a longitudinal descriptive study, the cases of four Portuguese HEIs that stood or stand out in the reporting of sustainability and, more recently, of the SDGs, are analysed: the FEUP; the UMinho; the UC and the ISCTE-UIL. Among others, these four HEIs signed, on October 31, 2019, the Letter of Intent, which establishes a commitment to sustainability principles and practices in Portuguese HEIs.

This analysis is based on secondary data, namely: public available reports (such as sustainability, SDG or annual reports) and the websites of the four HEIs, as well as the video recording [41] of a meeting organized by the Sustainable Campus Committee of the Energy for Sustainability (EfS) Initiative, held on March 23, 2021 titled 'Roundtable: Sustainability Report', with the participation of guest speakers (the Director for Sustainability – ISCTE - UIL; Commissioner for Sustainability - FEUP and Pro-Rector – UC Planning Department, EfS Initiative). The main purpose of this initiative was

Table 3.5 Portuguese HEIs in the GreenMetric Rankings

Year	World Ranking	University	Total Score	Setting and Infrastructure	Energy and Climate Change	Waste	Water	Transportation	Education
2020	89	University of Minho	7625	975	1375	1725	700	1250	1600
2020	172	Polytechnic Institute of Viana do Castelo	6975	1150	1700	975	650	1200	1300
2020	190	University of Aveiro	6875	725	1550	1575	600	925	1500
2020	386	ISCTE-University Institute of Lisbon	5625	350	1275	1350	350	1100	1200
2019	55	University of Minho	7575	975	1525	1650	700	1150	1575
2019	230	University of Aveiro	5900	550	1450	1200	525	1075	1100
2019	291	ISCTE-University Institute of Lisbon	5475	350	1200	1350	350	1100	1125
2019	355	Polytechnic Institute of Viana do Castelo	5175	975	1100	825	450	975	850
2018	68	University of Minho	7100	1025	1325	1500	700	1075	1475
2018	360	Polytechnic Institute of Viana do Castelo	4700	875	1225	300	100	1100	1100
2018	407	University of Aveiro	4450	500	1025	1050	300	650	925
2018	647	Polytechnic Institute of Santarem	2825	450	575	525	50	575	650
2017	48	University of Minho	6158	735	1383	1401	660	963	1016
2017	413	University of Aveiro	3923	618	694	1101	346	663	501
2017	539	Polytechnic Institute of Santarem	2969	686	534	774	30	411	534
2016	289	University of Coimbra	4162	988	791	876	375	699	433
2016	466	Polytechnic Institute of Santarem	2653	557	517	675	30	400	474
2015	272	University of Coimbra	3661	523	600	1050	525	450	513
2015	403	Polytechnic Institute of Santarem	1750	286	219	600	50	225	370
2014	193	University of Coimbra	5042	531	1060	1200	850	950	451
2014	359	Polytechnic Institute of Santarem	2299	328	450	675	0	450	396
2013	300	Polytechnic Institute of Santarem	2149						

Source: https://greenmetric.ui.ac.id/rankings/overall-rankings-2020, consulted in 6-7-2021.

to make a reflection on the practice, requisites and advantages of regular publication of sustainability reports by HEIs.

3.5.1 Faculty of Engineering of the University of Porto

FEUP is the largest faculty at the University of Porto. It currently has approximately 8,000 students, 1,500 employees (faculty members, researchers, scholarship holders and staff) and is organized into nine departments [41, 42]. Since 2009, FEUP has been a member of the Environmental Association for Universities and Colleges and, since 2020, a founding member of the 'Pacto para os Plásticos' (Plastics Pact) [43].

FEUP was a pioneer in the preparation of the Sustainability Report with its 2006 Sustainability Report. According to the Commissioner for Sustainability - FEUP [41], the challenge launched by the director of the Faculty required a benchmarking on the practices by foreign universities (United States, Canada, United Kingdom and Australia) that had embraced similar challenges, as well as on the appropriate tools to use. The GRI guidelines were the basis for the preparation of the first version of FEUP's Sustainability Report. Therefore, the structure of the report included the three dimensions of GRI sustainability, with the presentation of environmental, social and economic performance indicators [44].

However, as the GRI guidelines were not directly aimed at HEIs, there was a need to develop an appropriate methodology specific for them ([41], Commissioner for Sustainability – FEUP). For this purpose, a matrix was developed. The matrix incorporated the HEIs dimensions (academic community, governance, operations, education, research and impact on the community) correlated with sustainability dimensions (social, environmental and economic). Based on this matrix, the new version of the FEUP 2008 Sustainability Report presented a structure focused on four internal dimensions: academic community (characterization, working conditions and absenteeism, training, occupational safety on campus, safety and health on campus and wellness on campus); operations (environmental and economic indicators), education (admission to FEUP and students' performance) and FEUP's impact on society (economic and social impact) [45]. The same structure was replicated in the sustainability reports from 2009– 2012 (last Sustainability Report available on FEUP's website[4]).

[4]https://paginas.fe.up.pt/~sustent/estudos-e-relatorios/relatorios-sustentabilidade/, consulted in 6-7-2021.

In 2012, the website 'FEUP e Sustentabilidade' (FEUP and Sustainability) was developed to: raise awareness in the community towards more sustainable behaviours; inform about some of the most relevant indicators/initiatives to promote sustainability; give suggestions on more sustainable behaviours and encourage a culture of sustainability [46].

In 2015, the Commission for Sustainability was established, 'with the mission of contributing to a better society, incorporating the principles of sustainable development into FEUP's decision-making processes, reflecting on its core operations and activities (education, research and development and interaction with society)' [46]. The Commission developed the strategic plan for sustainability 2017–2037, whose strategic priorities include four strategic themes and 11 objectives. The themes are: Community; Teaching-Learning; Research and Development and Sustainable Campus. The objectives were selected taking into account the identified critical success factors, potential threats, the growing commitment to sustainability at the international level, as well as the 17 SDGs. The produced document lists the strategic objectives under each theme and describes, for some selected objectives, the 'strategic vision for 2037'. In the appendix, the themes, objectives and strategic sub-objectives, indicators and targets for 2037 are presented in a schematic way [46].

In the meantime, FEUP opted for discontinuing the publication of the sustainability report and decided to invest in the new 'FEUP and Sustainability' website[5] ([41], Commissioner for Sustainability - FEUP), developed in 2021, to communicate indicators and initiatives. The website is structured in nine themes: Commissionership; Education and Research; Sustainable Campus; Projects and Initiatives; Studies and Reports; News; Events; Partners and Contacts. On this website, FEUP highlights that its sustainability strategy is in line with the UN's 17 SDGs for 2030.

However, according to the Commissioner for Sustainability - FEUP [41], there is the objective of restarting the publication of the Sustainability Report as early as 2021, since this document is seen as an important accountability tool.

Regarding access to data and its possible automation for the preparation of the Sustainability Report, the Commissioner for Sustainability - FEUP [41] stressed that as for the sustainable campus, the technical services staff collect and transfer the information with no automation. Regarding the teaching and research contribution to the SDGs, as the information system is common to

[5] https://paginas.fe.up.pt/∼ sustent/, consulted in 6-7-2021.

the entire University of Porto, it is centralized and there is no autonomy to create new menus. This is a problematic area requiring authorization from the Rectory, which results in a time-consuming process.

3.5.2 University of Minho

Founded in 1973, the UMinho received its first students in the 1975/76 academic year. UMinho is a public foundation under private law, under the Legal Regime of HEIs. The UMinho is located in three different campi: one in the city of Braga and two others in the city of Guimarães. The UMinho is organised in 12 different Schools and Institutes: Schools of Architecture, Sciences, Medicine, Law, Economics and Management, Engineering, Psychology, Nursing and the Institutes of Social Sciences, Education and Arts, Human Sciences and Research Institute I3Bs. Currently, the UMinho has approximately 19,600 students, 13% of which are international students, 1,033 faculty members and 1,010 researchers and scholarships and 645 staff members [47].

The UMinho started its reporting on sustainability in 2010. This report covers the period from 2007– 2010 [48]. Subsequently, the Sustainability Reports for 2011, 2012–2013, 2014 and 2015 were published. On these reports, sustainability is assumed, in accordance with the Rector's message, as a strategic pillar of the university, in line with the Strategic Plan 2020 and the Programmatic Plan of 2013–17. These reports are available on the university's website[6].

The UMinho chose to follow the GRI guidelines. The 2010 and 2011 Sustainability Reports start by acknowledging the importance of SD for HEIs, followed by the presentation of UMinho and its governance structure. The reports proceed with the economic, social and environmental performance indicators disclosure [48, 49]. However, it is highlighted that the methodology is quite general, requiring some adjustment regarding its application to HEIs [48]. At the time, the GRI Guidelines considered three levels of application: C, B and A, which corresponded to sets of indicators of increasing number and complexity. For the 2010 and 2011 reports, the UMinho self-declared as level C [48, 49].

The 2012–2013, 2014 and 2015 Sustainability Reports were prepared according to the GRI guidelines, in the G4 version, corresponding to the highest level of coverage In Accordance – Comprehensive. The UMinho was

[6]https://www.uminho.pt/PT/uminho/Sustentabilidade/Paginas/default.aspx#/Documentos, consulted in 6-7-2021.

the 1st Portuguese university to report in accordance with the G4 guidelines, 2nd European university and 6th worldwide [50–52].

Due to the strong cultural activity of UMinho, a new dimension was introduced in these reports, the Cultural Dimension (relating to cultural heritage and events), presented after the Environmental, Social and Economic Dimensions. The information structure was adapted to the reality of the HEIs. The topics addressed in the reports were developed in accordance with the appreciation by the UMinho community, based on the AA1000 Accountability Principles Standard, considering the mission and strategy of the University: the teaching activity, research and interaction with society [50–52].

In the 2015 Sustainability Report, a reference is made to the SDGs and their role in the university, considering that 'Education, Research and Learning are at the core of the approaches to sustainable development' ([52], p. 13). This document presents the contribution of the University of Minho to the SDGs, as a result of the sustainability policies implemented since 2009 and stated in the Sustainability Reports and an appendix on the alignment with the SDGs.

The UMinho discontinued the publication of the Sustainability Report. However, the UMinho did not disregard its concerns with SD and sustainability, which are visible in its mission stated in the statutes of the university:

'The University's mission is to generate, disseminate and apply knowledge, based on freedom of thought and on the plurality of the critical exercises, promoting higher education and contributing to the construction of society model based on humanist principles, under which, knowledge, creativity and innovation are factors for growth, SD, well-being and solidarity' ([53], Part C, No. 115, p. 35).

The UMinho remained focused on demonstrating the results linked to its sustainability policy, supported by the 2017–2021 Action Plan [54], the 2020 Strategic Plan [55], the 2017 [56], 2018 [57] and 2019 [58] Annual Reports. The inclusion of the university in prestigious international rankings is also evidence of its commitment to sustainability.

In the annual reports, there are several references to the importance of the results obtained by UMinho in the UI GreenMetric World University Rankings. In 2017, UMinho was positioned as the 1st university in the country, 2nd in the Iberian Peninsula, 23rd in Europe and 48th in the world [56]. In the 2018 edition, UMinho was ranked 68th in the world, reinforcing again its 1st place in the country, 4th in the Iberian Peninsula and 33rd in

Europe [57]. In 2019, UMinho consolidated, for the third consecutive year, its leadership among Portuguese universities, appearing in the 55th position in the world and the 30th position among its European pairs. Also in 2019, UMinho was included, for the first time, in the THE Impact Rankings, having been distinguished with the 83rd position among 467 participants, conquering the 21st position in the SDG 4 - Quality Education [58].

Without an updated website focusing on sustainability (as the information contained therein refers to 2015) and without a recently published sustainability report (as the last one reports to 2015), UMinho innovated in the structure and content of the 2019 Annual Report. This HEI prepared and included, for the first time, a non-financial reporting in its annual report. This disclosure was made on a voluntary basis and following the best practices (see, for example, [33, 59, 60]), summarising information on public procurement practices, environmental and social performance:

'The need to disclose non-financial information is aimed not only for accountability purposes, (...) but also to disclose and explain the chosen policies followed in the development of the mission letters and the consequent creation of value in the short, medium and long term. Although there are no regulations to follow, the purpose of this disclosure is to invest in the introduction of performance measures that, progressively and in a convergent way, allow to move from separate disclosures, such as the annual publication of sustainability reports, to an integrated reporting, which enables a holistic view of the University of Minho, thus improving its reporting and accountability practices' ([58], p. 179).

3.5.3 University of Coimbra

The UC was founded in 1290. It currently includes ten teaching and research units (Faculty of Arts, Faculty of Law, Faculty of Medicine, Faculty of Science and Technology, Faculty of Pharmacy, Faculty of Economics, Faculty of Psychology and Educational Sciences, Faculty of Sport Sciences and Physical Education, Institute of Interdisciplinary Research and College of Arts), a research unit (Institute of Nuclear Sciences Applied to Health) and nine Cultural and Training Support Units (General Library, Archive, Press, Science Museum, 25th of April Documentation Centre, Gil Vicente Academic Theater, University Stadium, Health Sciences Library and Botanical Garden). The academic community has about 30,000 people (students, faculty and staff) [61].

The first Sustainability Report of the UC was produced by a multidisciplinary team, coordinated by four faculty members from civil engineering and the Faculty of Economics, for the 2019 year. The University's Planning Management Division, the Quality Promotion Office and the Building, Safety and Environment Management Service were involved in the preparation of the document, which is aligned with the UC's sustainability and SD strategy ([41], Pro-Rector - Planning Department of the University of Coimbra, Energy for Sustainability Initiative). Simultaneously, the UC prepared the first SDG Report for 2019/2020 [62]. These reports are available on the recent UC website dedicated to SD[7].

The 2019 Sustainability Report aims to assess and report the performance of the UC, in the various dimensions of sustainability and demonstrates the university's commitment to promoting sustainability and the SDGs, as reflected in its 2019–2023 Strategic Plan [61]. The report is structured into the following chapters: general characterization of the UC; UC governance; academic community and society (staff, faculty and community well-being); teaching and research; heritage and culture; environment, biodiversity and climate action; economy; and the GRI Tables. The inclusion of the chapter dedicated to heritage and culture is justified by the unique, historical, material and immaterial heritage of the UC that, with more than seven centuries, is part of the UNESCO World Heritage since 2013. The report was prepared based on the GRI guidelines (2016 version), under the essential option, adapted for HEIs. In addition to the Environmental, Social and Economic dimensions of sustainability reporting, the Education and Research and Culture and Heritage dimensions are also considered [61].

The 2019/2020 SDG Report highlights the classification obtained by the UC in the second edition of the THE Impact Rankings for 2020, namely, the 62nd position in the global ranking, among 766 universities worldwide and as the best Portuguese HEI in SDG 3 – Quality Health. The UC was also reported as the only Portuguese HEI in the worldwide Top 20 in the same SDG (17th position among 620 universities worldwide) (see table 3.3). The positions obtained by the UC in SDG 1, SDG 9 and SDG 14 are also highlighted [62].

Subsequently, the SDGs are presented in the Strategic Reference Framework of UC's 2019-2023 Strategic Plan and the UC contributions to the SDGs that are valid across all areas of SD. The UC adopted a bottom-up approach by carrying out surveys applied to a relevant part of the university's

[7]https://www.uc.pt/sustentabilidade, consulted in 6-7-2021.

'ecosystem' ([41], Pro-Rector - Planning Department of the University of Coimbra, Energy for Sustainability Initiative). The contributions of the scientific publications dimension to the SDGs were assessed using the THE Ranking methodology. In this assessment, the SDG 3, followed by SDG 7 and SDG 11, stood out. With regard to the contributions of educational offer dimension, all coordinators of 1st, 2nd and 3rd study cycles were asked to rate them using up to 3 SDGs and SDG 4, SDG 3 and SDG 9 stood out. To assess the contributions of the RandD units to SGDs, the questionnaires were applied to the directors of the research centers and SDG 3 and SDG 9 were highlighted As far as the research and innovation projects are concerned, researchers were surveyed about their research focus on the SDGs and again SDGs 3 and 9 were emphasized ([41], Pro-Rector - Planning Department of the University of Coimbra, Energy for Sustainability Initiative, [62]). Next, SDG Report presents the developed initiatives and activities for each SDG.

The Pro-Rector of the UC [41], manifested her concerns regarding the quality of sustainability reports to better reflect the HEIs reality. She classified the preparation process of such a report as challenging and stressful and, in the future, the UC will try to make new adjustments to better accommodate HEIs' reality.

Concerning the access and automation of data, which is particularly important to improve and speed up the process of producing the report, the rectory team is committed to working to promote and facilitate the process, by creating procedures and information flowcharts ([41], Pro-Rector - Planning Department of the University of Coimbra, Energy for Sustainability Initiative).

3.5.4 ISCTE-University Institute of Lisbon

Created in 1972, ISCTE-UIL is a public university that in 2009 was established as a foundation. Its four Schools (ECSH - School of Social and Human Sciences; ESPP - School of Sociology and Public Policy; ISTA - School of Technologies and Architecture; IBS - Business School), 16 Departments and eight Research Units share the same campus in the center of Lisbon. Currently, ISCTE has about 10,000 students in undergraduate (44%) and graduate (56%) programs, 401 faculty members (ETI), 415 researchers at 100% and 259 staff members [63].

The first reflection on sustainability arose in 2012 with the creation of a group dedicated to university social responsibility [63]. Despite these initial concerns, according to the Director for Sustainability-ISCTE-UIL [41], only in 2016 the ISCTE-UIL's sustainability team began to work in a more structured way. In that year, a set of actions to improve sustainable performance were set under the Sustainability@ISCTE-IUL project [63]. The work carried out under the environmental pillar led ISCTE-UIL to be the first Portuguese university receive the environmental certification ISO 14001:2015, in 2018. In the following year, the ISCTE-IUL started the implementation of a social responsibility management system aiming at the Corporate Social Responsibility certification in 2020 [63].

These initiatives under the Sustainability@ISCTE-IUL project set up the foundations for the preparation of the Sustainability Report. As described by the Director for Sustainability – ISCTE-UIL [41], a consulting company was hired to help to develop the sustainability report for the years 2018 and 2019. This company provided technical assistance in defining the content of the report, collecting and managing the information. The process started with a brainstorming activity that led to the definition of the structure and topics to include in the report. The identified topics are: Sustainability of the supply chain; Sustainability governance; Communication and involvement with stakeholders; Social support; Diversity and inclusion; Training and career management; Health, safety and well-being; Energy and climate change; Use and management of natural resources; Sustainable events; Teaching-learning; Research; Interaction with society [63]. In the following months, data, to fulfil the information needs of each topic, was requested to the different members of the ISCTE-UIL community (students' association, pedagogical council, scientific council, research centres', departments, schools, social action, human resources) ([41], Director for Sustainability – ISCTE-UIL).

The Director for Sustainability – ISCTE-UIL [41] described the process as a very dynamic one, but also as very time-consuming. While the Sustainability team was small, the top management of ISCTE-IUL was involved (Vice-Rector), as well as approximately 30 different stakeholders who promoted the communication of the sustainability message, encouraged quality and sustainability practices and participated in monitoring and auditing activities [see also 63]. All gathered information was compiled in August 2020 and ISCTE-UIL's first Sustainability Report was published on November 12, 2020, on World Quality Day ([41], Director for Sustainability – ISCTE-UIL).

The ISCTE-IUL's Sustainability Report uses GRI guidelines to communicate the performance of this HEI in different dimensions: Planet, People and

Prosperity. It also recognises its role in promoting sustainability, in accordance with the SDGs [64]. The ISCTE-UIL Sustainability Report starts with an 'Overview', then follows the sections: 'Who we are and what we do', 'Our approach to sustainability', 'Teaching-learning', 'Interaction with society', 'Sustainable campus', 'About this report' and finally, the Appendices. There is a visible focus on the SDGs throughout the report. The Appendices present the correspondence table, the SDG table and the GRI table.

GRI indicators were used for the operations/environmental management system, while for teaching-learning and research activities other specific indicators were created ([41], Director for Sustainability – ISCTE-UIL). The structure of chapters and sub-chapters corresponds to the 13 previously defined topics, with the main SDGs identified for each topic [63]. ISCTE-UIL was the first national HEI to classify its activities according to the SDGs. Course units are rated up to a maximum of 3 SDGs by coordinators, as are dissertations by students, research by the authors themselves and research projects, reflecting the involvement of all. The classifications obtained by ISCTE-UIL in the rankings are also highlighted, namely in the world's top 300 and 4th in Portugal in THE Impact Rankings (2019 Ed.) regarding its performance in the implementation of the SDGs, and in 3rd place at the national level and 291st internationally in 2019 in the GreenMetric World University Rankings, which ranks the most sustainable universities [63].

The report incorporates visual images with the narratives and has a modern and appealing design. Its digital version [63] is interactive, with links to various initiatives, themes and other documents, for a better understanding of each subject. The Director for Sustainability – ISCTE-UIL [41] highlighted the positive feedback she received from various stakeholders. Regarding the access to data and its automation for the preparation of the Sustainability Report, the Director for Sustainability – ISCTE-UIL [41] highlighted that the objective is to take advantage of existing reporting structures, whenever possible. However, despite the easier access to data on consumption (e.g. water, gas, electricity), nothing is truly automated. Also, information disclosed on residuals is based on estimates. In what relates to teaching and research and their contributions to the SDGs, the questions were asked directly in the information systems that already exist in ISCTE-IUL, which eased the collection of information, and the objective is to take advantage of existing reporting structures whenever possible.

3.6 Conclusion

The Sustainable Development paradigm represents a significant shift from the previous traditional economic development paradigm. The challenge is to prospect a future, pursuing development and improved life quality, balancing environmental, social and economic aspects.

In comparison to its predecessor - the MDGs -, the 2030 Agenda is much more ambitious comprising 17 goals, 169 indicators and 223 targets cutting across economic, social and environmental dimensions, indicating five areas of critical importance: People, Planet, Peace, Prosperity and Partnerships [6].

Although the implementation of SDGs is a common responsibility for different stakeholders in society, HEIs are expected to play a leading role as a driving force behind the necessary transformations to be implemented in societies, actively contributing towards a truly new paradigm shift. Nonetheless the efforts following the 1st Sustainable Campus Conference (CCS 2019), Portuguese HEIs need to speed up to match the 2030 Agenda, both in terms of ambition and scale.

The participation of HEIs in sustainability rankings can be helpful to set long-term sustainability goals and priorities. Worldwide, the sustainability competition is rising as the number of HEIs included in sustainability rankings is increasing on an annual basis. This puts more pressure on Portuguese HEIs, as an improved score might not result in a better worldwide ranking position. Nevertheless, at this level some progress was achieved in recent years and currently, the UMinho, the University of Aveiro and the ISCTE-UIL own the best positions among Portuguese HEIs in important rankings like GreenMetric World Universities Rankings and THE Impact Rankings.

HEIs commitment to SD can also be strengthened through sustainability reporting. In fact, sustainability reporting can help HEIs to measure, understand and communicate their performance along with different scopes like economic, environmental and social dimensions. It also helps HEIs to set their specific goals and manage change more effectively. The cases of four Portuguese HEIs committed to sustainability principles and practices as signatories of the Letter of Intent, namely FEUP, UMinho, UC and ISCTE-UIL, were analysed. Despite not being aimed at HEIs, GRI guidelines are generally adjusted and used for sustainability reporting purposes by the four institutions. In general, these Portuguese HEIs consider the process of preparing the sustainability report as complex, because it involves a multitude of dimensions and a significant amount of information, and is very time-consuming. The automation of the process is pointed out as a mitigating factor for this problem. Ultimately, these Portuguese HEIs acknowledge the fact that sustainability reporting is very important as it enhances accountability and transparency and is working towards the improvement of their reporting processes.

While this chapter is being written, the world is experiencing the unexpected global COVID-19 pandemic and the associated global economic crisis. This crisis is contributing to the severe deterioration of the living and health conditions of millions of people worldwide. In fact, 'With up to 100 million more people being pushed into extreme poverty in 2020, 1.4 billion children affected by school closures and more than 400 thousand confirmed deaths from COVID-19 as of early June, the pandemic is hitting all human development dimensions hard, in all countries, almost at the same time' ([65], p. 3).

The economic and financial crisis resulting from the pandemic situation constitutes a major pressure element for the 2030 Agenda as the focus of the states and citizens tends to be on economic survival instead of being in building a better sustainable future. Although the COVID-19 health crisis has caused deep social and economic disruptions, it can also be seen as an opportunity to accelerate demanded changes under 2030 Agenda: '(...) Governments and societies face unprecedented policy, regulatory and fiscal choices as they act to save lives and set a course for a sustainable future. The choices made today, if made well, could be the tipping points that transform our societies and our planet for the better' ([65], p. 3).

HEIs serve a wide variety of stakeholders through their functions of education, research and service. Therefore, HEIs are in an exceptional position to react to the current disruptive and challenging times by enhancing economic recovery through social cohesion and innovation. As far as Portuguese HEIs are concerned, the current health and economic crisis might work as an inflective point for them to definitely accelerate sustainability and its contribution to the achievement of the SDGs by, not only create knowledge through research but also by develop skills and inspire action of other societal stakeholders.

References

[1] WCED (1987). Our common future. Report of the world commission on environment and development: our common future. Available at: https://sustainabledevelopment.un.org/content/documents/5987our-common-future.pdf, last accessed July 12, 2021.

[2] Mensah, J. (2019). Sustainable development: Meaning, history, principles, pillars, and implications for human action: Literature review. *Cogent Social Sciences*, vol. 5, no. 1, pp. 1653531.

[3] Waas, T., Hugé, J., Verbruggen, A. and Wright, T. (2011). Sustainable development: a bird's eye view, *Sustainability*, vol. 3, no. 12, pp. 1637–1661.

[4] Porter, M. E., and van der Linde, C. (1995). Toward a new conception of the environment competitiveness relationship. *Journal of Economic Perspectives*, vol. 9, pp. 97–118.

[5] United Nations (UN) (2015). The Millennium development goals report 2015, replace by: available at: https://www.un.org/millenniumgoals/2015_MDG_Report/pdf/MDG%202015%20rev%20(July%201).pdf, last accessed December 7, 2021.

[6] United Nations UN (2015). Transforming our world: The 2030 agenda for sustainable development. New York: United Nations, Department of Economic and Social Affairs. Available at: https://sdgs.un.org/sites/default/files/publications/21252030%20Agenda%20for%20Sustainable%20Development%20web.pdf, last accessed December 7, 2021.

[7] UNESCO (2014). Shaping the Future We Want. UN Decade of Education for Sustainable Development (2005–2004). Final Report. DESD Monitoring and Evaluation. United Nations Educational, Scientific and Cultural Organization, France.

[8] Abad-Segura, E., and González-Zamar, M. D. (2021). Sustainable economic development in higher education institutions: A global analysis within the SDGs framework. *Journal of Cleaner Production*, vol. 294, pp. 126–133.

[9] Albareda-Tiana, S., Vidal-Raméntol, S., and Fernández-Morilla, M. (2018). Implementing the sustainable development goals at University level. *International Journal of Sustainability in Higher Education*, vol. 19, no. 3, pp. 473–497.

[10] Cortese, A.D. (2003). The critical role of higher education in creating a sustainable future, *Planning for Higher Education*, vol. 31, pp. 15–22.

[11] Lozano, R. (2011). The state of sustainability reporting in universities, *International Journal of Sustainability in Higher Education*, vol. 12, no. 1, pp. 67–78.

[12] Arroyo, P. (2017). A new taxonomy for examining the multi-role of campus sustainability assessments in organizational change. *Journal of Cleaner Production*, vol. 140, pp. 1763–1774.

[13] Lozano, R., Ceulemans, K., Alonso-Almeida, M., Huisingh, D., Lozano, F.J., Waas, T., Lambrechts, W., Lukman, R. and Hug, J. (2015). A review of commitment and implementation of sustainable development in higher education: results from a worldwide survey, *Journal of Cleaner Production*, vol. 108, pp. 1–18.

[14] Lozano, R. (2006). Incorporation and institutionalization of SD into universities: breaking through barriers to change, *Journal of Cleaner Production*, vol. 14, no. (9/11), pp. 787–796.

[15] Lozano, R., Lukman, R., Lozano, F. J., Huisingh, D., and Lambrechts, W. (2013). Declarations for sustainability in higher education: becoming better leaders, through addressing the university system. *Journal of Cleaner Production*, vol. 48, pp. 10–19.

[16] SDSN Australia/Pacific (2017): Getting started with the SDGs in universities: A guide for universities, higher education institutions, and the academic sector. Australia, New Zealand and Pacific Edition. Sustainable Development Solutions Network – Australia/Pacific, Melbourne. Available at: https://ap-unsdsn.org/wp-content/uploads/University-SDG-Guide_web.pdf, last accessed December 7, 2021.

[17] Rosen, M. A. (2019). Do universities contribute to sustainable development?. *European Journal of Sustainable Development Research*, vol. 4, no. 2, pp. em0112.

[18] Sonetti, G., Brown, M., and Naboni, E. (2019). About the triggering of UN sustainable development goals and regenerative sustainability in higher education. *Sustainability*, vol. 11, no. 1, pp. 254.

[19] Gadotti, M. (2008), What we need to learn to save the planet, *Journal of Education for Sustainable Development*, vol. 2, no. 1, pp. 21–30.

[20] Gomes, S. F., Jorge, S., and Eugénio, T. P. (2020). Teaching sustainable development in business sciences degrees: evidence from Portugal. *Sustainability Accounting, Management and Policy Journal*, vol. 12, no. 3, pp. 611–634.

[21] Farinha, C. S.; Azeiteiro, U. and Caeiro, S.S. (2018). Education for sustainable development in Portuguese universities: The key actors'opinions. *International Journal of Sustainability in Higher Education*, vol. 19, no. 5, pp. 912–941.

[22] Rosen, M. A. (2018). Issues, concepts and applications for sustainability. *Glocalism: Journal of Culture, Politics and Innovation*, vol. 3, pp. 1–21.

[23] Schneider, F., Kläy, A., Zimmermann, A. B., Buser, T., Ingalls, M., and Messerli, P. (2019). How can science support the 2030 Agenda for

Sustainable Development? Four tasks to tackle the normative dimension of sustainability. *Sustainability Science*, vol. 14, no. 6, pp. 1593–1604.

[24] Leal Filho, W., Shiel, C., do Paço, A., and Brandli, L. (2015). Putting sustainable development in practice: Campus greening as a tool for institutional sustainability efforts. In Sustainability in higher education (pp. 1–19). Chandos Publishing.

[25] Ramísio, P. J., Pinto, L. M. C., Gouveia, N., Costa, H., and Arezes, D. (2019). Sustainability strategy in higher education institutions: Lessons learned from a nine-year case study. *Journal of Cleaner Production*, vol. 222, pp. 300–309.

[26] Rede Campus Sustentável (2021). Declarações para implementação da sustentabilidade nas IES, available at: www.redecampussustentavel.pt/declaracoes-de-sustentabilidade/, last accessed July 12, 2021.

[27] Alghamdi, N.; den Heijer, A.; de Jonge, H. (2017). Assessment tools' indicators for sustainability in universities: An analytical overview. *International Journal of Sustainability Higher Education*, vol. 18, pp. 84–115.

[28] Caeiro, S.; Sandoval Hamón, L.A.; Martins, R.; Bayas Aldaz, C.E. (2020). Sustainability assessment and benchmarking in higher education institutions—A Critical Reflection. *Sustainability*, vol. 12, pp. 543.

[29] Findler, F.; Schönherr, N.; Lozano, R.; Stacherl, B. (2018). Assessing the Impacts of Higher Education Institutions on Sustainable Development—An Analysis of Tools and Indicators. *Sustainability*, Vol. 11, no. 1, pp. 59.

[30] Ramos, T.; Pires, M.S. (2013). Sustainability assessment: The role of indicators. In Sustainability Assessment Tools in Higher Education—Mapping Trends and Good Practices at Universities around the World; Caeiro, S., Leal Filho, W., Jabbour, C., Azeiteiro, U., Eds.; Springer: Cham, Switzerland: pp. 81–100.

[31] GRI (n.d.). About GRI, available at: https://www.globalreporting.org/about-gri/, last accessed on July 12, 2021.

[32] Larrán Jorge, M., Andrades Peña, F. J., and Herrera Madueño, J. (2019). An analysis of university sustainability reports from the GRI database: an examination of influential variables. *Journal of Environmental Planning and Management*, vol. 62, no. 6, pp. 1019–1044.

[33] International Integrated Reporting Council (IIRC) (2021). International IR framework, available at https://integratedreporting.org/wp-content/uploads/2021/01/InternationalIntegratedReportingFramework.pdf, last accessed December 7, 2021.

[34] Comissão Nacional da UNESCO – Portugal (2006). Década das Nações Unidas da Educação para o Desenvolvimento Sustentável (2005–2014). Contributos para a sua dinamização em Portugal. Retirado de www.dge.mec.pt/sites/default/files/ECidadania/Areas_Tematicas/contibutos_dnuds.pdf, last accessed July 6, 2021.

[35] Aleixo, A.M., Azeiteiro, U.M. and Leal, S. (2016). "Toward sustainability through higher education: sustainable development incorporation into Portuguese higher education institutions", in Davim, J.P. and Leal Filho, W. (Eds), *Challenges in Higher Education for Sustainability*, Springer, London: pp. 159–187.

[36] Farinha, C.; Caeiro, S. and Azeiteiro, U. (2019). Sustainability strategies in Portuguese Higher Education Institutions: Commitments and practices from internal insights. *Sustainability*. vol. 11, no. (3227), pp. 1–25.

[37] Matos, A.; Cabo, P.; Ribeiro, M. and Fernandes, A. (2015). As instituições de Ensino Superior perante a problemática ambiental (The Higher Education institutions in relation to environmental problems). EDUSER: Revista de Educação. vol. 7, no. 2, pp. 13–40.

[38] Aleixo, A.M., Azeiteiro, U. and Leal, S. (2018). The implementation of sustainability practices in Portuguese higher education institution. *International Journal of Sustainability in Higher Education*. vol. 19, no. 1, pp. 146–178.

[39] Rede Campus Sustentável - Portugal (2021). Carta de intenções – Compromisso das Instituições de Ensino Superior com o desenvolvimento sustentável. Available at www.redecampussustentavel.pt/carta-de-intencoes, last accessed July 6, 2021.

[40] The Sustainability Tracking, Assessment & Rating System (2019). IN-13: Spend Analysis-Universidade Aberta. Available at: https://reports.aashe.org/institutions/universidade-aberta-/report/2019-04-05/, last accessed on Jul. 7, 2021.

[41] Sustainable Campus Committee of the Energy for Sustainability (EfS) Initiative (2021). "Roundtable: Sustainability Report", held on Mar. 23, 2021, available at www.youtube.com/watch?v=2WSKG-OmZ_g, last accessed on Jul. 7, 2021.

[42] Faculdade de Engenharia Universidade do Porto (FEUP) (2021). FEUP em Números, available at: https://sigarra.up.pt/feup/pt/web_base.gera_pagina?p_pagina=269649, last accessed on Jul. 7, 2021.

[43] Faculdade de Engenharia Universidade do Porto (FEUP) (2021). FEUP e Sustentabilidade, available at: https://paginas.fe.up.pt/~sustent/, last accessed on Jul. 7, 2021.

[44] Faculdade de Engenharia Universidade do Porto (FEUP) (2006). Relatório de Sustentabilidade 2006, available at https://paginas.fe.up.pt/~sustent/estudos-e-relatorios/relatorios-sustentabilidade/, last accessed on Jul. 7, 2021.

[45] Faculdade de Engenharia Universidade do Porto (FEUP) (2008). Relatório de Sustentabilidade 2008, available at https://paginas.fe.up.pt/~sustent/estudos-e-relatorios/relatorios-sustentabilidade/, last accessed on Jul. 7, 2021.

[46] Faculdade de Engenharia Universidade do Porto (FEUP) (2017). Plano Estratégico 2017-2037, available at https://paginas.fe.up.pt/~sustent/wp-content/uploads/2020/05/20180503_Estrategia_Sustentabilidade.pdf, last accessed on Jul. 7, 2021.

[47] Universidade do Minho (2021). Facts and Figures, available at www.uminho.pt/EN/uminho/Pages/facts-and-figures.aspx, last accessed on Jul. 7, 2021.

[48] Universidade do Minho (2012). Relatório de Sustentabilidade Ano de 2010, available at www.uminho.pt/PT/uminho/Informacao-Institucional/RelatoriosSustentabilidade/relatorio-de-sustentabilidade-2010.pdf, last accessed on Jul. 12, 2021.

[49] Universidade do Minho (2013). Relatório de Sustentabilidade Ano de 2011, available at www.uminho.pt/PT/uminho/Informacao-Institucional/RelatoriosSustentabilidade/relatorio-de-sustentabilidade-2011-af.pdf, last accessed on Jul. 12, 2021.

[50] Universidade do Minho (2014). Relatório de Sustentabilidade Ano de 2012-13, available at www.uminho.pt/PT/uminho/Informacao-Institucional/RelatoriosSustentabilidade/relatorio_sustentabilidade_2012_2013.pdf, last accessed on Jul. 12, 2021.

[51] Universidade do Minho (2015). Relatório de Sustentabilidade Ano de 2014, available at https://www.uminho.pt/PT/uminho/Informacao-Institucional/RelatoriosSustentabilidade/relatorio_sustentabilidade_2014.pdf, last accessed on Jul. 12, 2021.

[52] Universidade do Minho (2016). Relatório de Sustentabilidade Ano de 2015, available at https://www.uminho.pt/PT/uminho/Informacao-Institucional/RelatoriosSustentabilidade/Relatorio%20de%20Sustentablidade%202015.pdf, last accessed on Jul. 12, 2021.

[53] Estatutos da Universidade do Minho (2021), Normative Order No. 15-2021 published in the Official Gazette, Diário da República, 2nd Series, of June 24th, 2021, available at https://www.uminho.pt/PT/uminho/Informacao-Institucional/Paginas/Estatutos.aspx, last accessed on Jul. 12, 2021.
[54] Universidade do Minho (2018). Plano de ação para o quadriénio 2017–2021, available at www.uminho.pt/PT/uminho/Informacao-Institucional/Planos-e-Relatorios/3UltimosPQuad/Plano-de-ac%CC%A7a%CC%83o-2017-2021.pdf, last access on Jul. 12, 2021.
[55] Universidade do Minho (2013). Stategic Plan 2020, available at www.uminho.pt/PT/uminho/Informacao-Institucional/Planos-e-Relatorios/PlanoEstrategico/UMINHO-2020.pdf, last access on Jul. 12, 2021.
[56] Universidade do Minho (2017). Relatório de Gestão e Contas Consolidadas, available at: https://www.uminho.pt/PT/uminho/Informacao-Institucional/Planos-e-Relatorios/3UltimosRAtividadesUMinho/Relatorio_Gestao_Contas_Consolidadas_2017.pdf, last accessed Jul. 12, 2021.
[57] Universidade do Minho (2018). Relatório de Atividades e Contas 2018, available at www.uminho.pt/PT/uminho/Informacao-Institucional/Planos-e-Relatorios/RelatoriosAtividadesContas/Relato%CC%81rio_Geral_de_Contas_Individual_e_Consolidado_2018_UMinho.pdf, last access on Jul. 12, 2021.
[58] Universidade do Minho (2019). Relatório de Atividades e Contas Separadas 2019, available at, www.uminho.pt/PT/uminho/Informacao-Institucional/Planos-e-Relatorios/RelatoriosAtividadesContas/Relatorio_Atividades_Contas%20Separadas_UMinho2019.pdf, last access on Jul. 12, 2021.
[59] European Union (2014). Directive 2014/95/EU-Disclosure by large companies and non-financial information groups and information on diversity, OJ L 330, 15.11.2014, p. 1–9 available at https://eur-lex.europa.eu/legal-content/EN/TXT/?uri=CELEX%3A32014L0095, last accessed December 7, 2021.
[60] Decree-Law 89/2017, published in the Official Gazette, Diário da República n.145/2017, Serie I, of Jul. 28th, 2017, available at https://dre.pt/application/conteudo/107773645, last accessed December 7, 2021.
[61] Universidade de Coimbra (2020). Relatório de Sustentabilidade da UC 2019, available at https://www.uc.pt/sustentabilidade/RelatorioSustentabilidade2019_dup_web.pdf, last accessed on Jul. 9, 2021.

[62] Universidade de Coimbra (2020). Relatório ODS 2019-20, available at www.uc.pt/sustentabilidade/UC_construir_mundo_diferente_relatorio_ODS2019_2020.pdf, last accessed on Jul. 9, 2021.
[63] ISCTE-IUL (2020). Relatório de Sustentabilidade 2018/19, available at www.iscte-iul.pt/assets/files/2020/11/11/1605098779545_Rel_Sust_2019_ISCTE.pdf, last accessed on Jul. 9, 2021.
[64] ISCTE-IUL (2020). Iscte publica pela 1ª vez Relatório de Sustentabilidade, available at www.iscte-iul.pt/noticias/1740/iscte-publica-pela-1-relatorio-de-sustentabilidade, last accessed on Jul. 9, 2021.
[65] United Nations Development Programme (UNDP) (2020). Beyond Recovery: Towards 2030, available at: https://www.undp.org/publications/beyond-recovery-towards-2030#modal-publication-download, last accessed December 7, 2021.

4

Where is the Brazilian Higher Education Within the Sustainable Development Goal 4?

Sidney L. M. Mello[1,2,*], Carlos E. Bielschowsky[3], Marcelo J. Meriño[1], and Thaís N. da R. Sampaio[1]

[1]Universidade Federal Fluminense, Escola de Engenharia, Laboratório de Tecnologia e Gestão de Negócios, Niterói, RJ, 24210-240, Brasil
[2]Faculdade Cesgranrio, Rio de Janeiro, RJ, 22241-125, Brasil
[3]Universidade Federal do Rio de Janeiro, Centro de Tecnologia, Instituto de Química, Cidade Universitária, Rio de Janeiro, RJ, 21941-909, Brasil
E-mail: smello@id.uff.br
*Corresponding Author

Abstract

Higher education (HE) in Brazil has always been a privilege of the elites. In the 2000s, this scenario changes with a global trend towards inclusion in education as a necessity for social and economic transformations. In 2015, the UN 2030 agenda defined HE as a key objective for the planet's sustainable development. Brazil embraced the agenda with the challenges of resolving social inequalities by promoting equal access to tertiary education. The need for technical and scientific qualification of the young population is critical for the development in aspects such as income, employment, and social justice. Between 1999 and 2019, HE grew 263%, from 2.37 million enrollments in 1999 to 8.60 million in 2019. The private sector grew faster than the public, and distance learning enrollments have risen sharply in the last 10 years at an average rate of about 11%. This growth fits principles of sustainable development goal 4 (SDG 4), which aims to train young people and adults in professional skills to achieve decent work and employment. Brazil has kept an educational policy in full swing with reforms necessary to place the country in the world of work in the 21st century. But it faces substantial issues on oligopolies and loss of quality of private HE with 80% of total enrollments. The current Brazilian education plan needs to solve the quality assessment of the private sector on-campus and distance learning that grows very fast. Further, it needs to increase enrollments in the public sector, which is highly qualified.

Keywords: Sustainable development, SDG 4, distance learning; Brazilian higher education; education policy.

4.1 Introduction

Meeting the needs of the current generation without compromising the ability of future generations to meet their own needs is key to the concept of sustainable development, defined from the Brundtland Report in the document entitled Our Common Future [1, 2]. The report refers to the need to empower people, now and in the future, to achieve a satisfactory level of human, social, and economic development and cultural achievement while making reasonable use of natural resources and preserving species and habitats.

The sustainable development concept served as background for the events of the Millennium Summit, which culminated with the Millennium Declaration and eight global development goals until 2015 [3]. Quality primary education for all prominently appeared among the Millennium Development Goals (MDGs), much inspired by the Education for All (EFA) movement conceived during the World Education Forum in Dakar [4].

At the time, HE did not figure as a priority on global political agendas, apparently due to an understanding that developing countries would obtain a greater return by investing financial resources in primary and secondary education instead of universities or technical training [5]. Notably, a narrow view on the role of HE in society, whose responsibility is the training of teachers, professionals, qualified technical workers and promoting advances in scientific research, innovation, and technology.

Despite this, HE has regionally and globally adopted its reform agenda to solve the growing demand for qualified labor needed for socio-economic development. HE has become a necessity for developed and developing countries to achieve sustainable development [6]. Furthermore, advances in the knowledge society and the globalization of the economy have contributed significantly to pressure access to HE [7]. The total number of enrollments in

HE increased from 146 million in 2006 to more than 218 million students in 2016 and is expected to reach, according to estimates, more than 265 million in 2025 [8].

In 2015, the UN Agenda 2030 defined an action plan for the planet's sustainable development and, finally, formalized HE on the table of global development policies [9]. Among 17 interdependent sustainable development goals (SDGs), the SDG 4 addresses educational challenges, including HE [10, 11]. Such a goal recommends that inclusive, equitable, and quality education must be guaranteed as a learning opportunity throughout life for all. The targets of the SDG 4 specifically address the following: (1) ensuring equal access for all women and men at affordable prices to quality technical and vocational education, including university education; (2) ensuring a substantial increase in the number of young people and adults with relevant skills, including technical and professional skills, for jobs, decent work, and entrepreneurship; (3) promote the offer of scholarships for international student mobility and; (4) challenge universities to develop key sustainability concepts in the course curriculum, involving climate change, human rights, and peace studies.

Enrollments in HE continue to grow worldwide in the wake of technological innovations, the best qualification for the job market, and social well-being. Massive access to HE has become an imperative for all countries and has earned the label of mass HE or massification. It means that HE is no longer exclusive to elites and embraces social groups that have been excluded from this possibility [12]. Among the many reasons for the increase in enrollment rates are the population growth, the need for better-trained professionals for competitive jobs, and a set of government policies oriented towards sustainable development, allowing women, ethnic and social minorities equal access to HE [13].

The main driving force behind the increase in demand for university degrees has been from the lower middle and middle classes. Governments have responded by finding ways to support the increase in enrollment, building new educational models, and more university and non-university institutions [14]. High rates of high school completion also have heated enrollment in preparing for new careers, professions, and life opportunities. With the increase in demand, registrations in the private sector have grown continuously, representing 30% of all international enrollments [15]. In Latin America, the private sector holds 49% of total registrations, while in Asia, they reach 36% on average [16]. New private providers have also emerged in university campuses with international branches and international online institutions [17]. Instead of research programs or other types of specialized studies, new institutions focus on opening as many places as possible for university students.

Another factor is the enrollment of non-traditional students in HE, including part-time students and working adults. Adults over 25 represent more than a third of undergraduate students enrolled in ten European countries, while in five countries, at least one in four students has part-time status [18]. Distance learning providers are gaining ground in tertiary education. The growth of distance learning has been global, with around 21.3% of enrollments of HE students worldwide [19]. Gross enrollment rates in Turkey grew from 30% in 2004–86% in 2014, partly due to distance education enrollments [20]. China has also used distance learning to transform HE into a mass system [21].

However, there is a lot of criticism concerning mass education [22]. An increasing stratification accompanies equity in access in social groups, quality, and prestige of the institutions. For this reason, there is disagreement as to whether there is justice in this process [23]. For [24], there is a functionalist concept in massification regarding putting the right people in the correct social positions for the general benefit of all. Some authors also think that low-income students are restricted to low-quality education [14]. For others, massification does not represent a democratization of education but a commodification of HE [23] or a university credential that reproduces inequalities in the labor market [25].

[26] considered that massification does not end inequalities but can promote advances resulting from the expansion of HE systems, which means the following: (1) provides more significant opportunities for social mobility; (2) promotes higher levels of income for the population; and (3) paves the way for the inclusion of women and historically marginalized social groups around the world. For [27], the expansion of HE cannot be separated from social and economic policies. Also, the SDGs are a universal call against poverty and a commitment to protecting the planet to ensure that all people have education [28, 29].

Since the 1990s, the traditional configuration of Brazilian HE systems has changed with the creation of various university and non-university institutions, public or private, both on-campus and distance learning. These reforms were inspired by global policies aimed at sustainable development, focusing on overcoming educational inequalities, expanding enrollment in HE, and offering vocational training [30, 31]. Economic support policies for low-income students and affirmative actions, such as ethnic quotas, were triggered to enable equitable access to HE [32].

For [33], HE in Brazil expanded with improved access to quality education and scientific production. The best public universities research-driven have become more inclusive of low-income students through the National Student Assistance Plan (Pnaes) [34] and the ethnic quota law [35, 36]. Private HE institutions, on-campus, and distance education together reached over 75% of enrollments, reflecting a global trend in which the public sector cannot support mass or universal HE. For this reason, the Brazilian government created public programs, such as the University for All Program (ProUni) and the HE Student Financing Fund (Fies), to subsidize low-income students in private institutions [37, 38]. Further, there has been notable growth in distance learning (DL) enrollments across the country. [39] showed that DL registered 49 thousand registrations in 2003 and almost 1.5 million in 2016. Between 2010 and 2015, enrollments in undergraduate courses grew 25.3% in total; on-campus enrollments grew by 21.1%, but DL enrollments grew by 49.7% [40]. In 2019, the DL enrollments reached 2.45 million undergraduate students [41].

This chapter deals with the advances and challenges of HE in Brazil and its alignment with SDG 4 for expanding access to the university degree and post-secondary professional training. It shows how, since the 1990s, HE has evolved in terms of enrollment looking for mass education, including equity access, professional training, and quality education. It also highlights how DL stands out in meeting the goals of registration in HE. Finally, it offers a critical view of the present and a perspective for the future, discussing the challenges concerning quality and full compliance with SDG 4 within the sustainable development in the country.

4.2 Expansion of HE in Brazil

The expansion of enrollments in HE in Brazil started, in fact, with the National Education Plan (PNE) action for the period between 2001 and 2011 [42]. With a considerable delay, this plan met the urgent need for reforms in education in the country due to the increasing pressure from high school graduates and a specialized workforce. There was also a perspective to follow a global trend on the massification and internationalization of HE as essential guidelines for economic globalization. To this end, the HE system should create a diverse set of institutions to meet different demands and social functions. A turning point in educational policies from elites to mass and universal HE.

The PNE (2001—2011) was created with the idea that knowledge is the basis of scientific and technological development, making the dynamism of current societies. Higher education institutions (HEIs) should be responsible for raising Brazil to the demands and challenges of the 21st century, finding the solution to current problems in all fields of life and human activity for a better future, reducing inequalities. Also, the provision of quality primary education for all was deposited in the hands of these institutions, as they form the teaching profession professionals; technical, scientific, and cultural professional staff of a higher level, the production of research and innovation capable of projecting the development of the country. In 2014, Brazil renewed such principles for a new PNE (2014–2024) as an even more pronounced focus on HE [43]. The current PNE is committed to the targets of SDG 4, explicitly contributing to the universalization of education, equity, and quality at different educational levels. One of the commitments of the PNE's goal 12 is to increase the gross enrollment rate in HE to 50% among the population aged 18–24, with the expansion of the public system to 40% of new enrollments. This goal will require the registration of about 11 million people in HE; according to projections that in 2024 the Brazilian population aged 18–24 will be 22.1 million.

Indeed, since 1999, efforts have been made to expand federal funds for education [44]. In 2018, it was invested 6% of the Brazilian Gross Domestic Product (GDP), about $113.1 billion. Thus, new public and private HE institutions have been created across the country, focusing on achieving the goals of the national plans, promoting inclusion and equity in access to HE. Between 1999 and 2019, HE in Brazil grew 263%, from 2.37 million enrollments in 1999 to 8.60 million in 2019. Between 1999 and 2006, the private sector grew somewhat faster than the public, rising from 65% of total enrollments to 75%; as of 2006, both sectors, public and private, grew proportionately, maintaining the rate of one–three, respectively (Figure 4.1).

The increase in the number of student enrollments at federal institutions resulted from the Program for Restructuring and Expansion Plans at Federal Universities (Reuni), the expansion of HE programs in the Technical Schools and the DL National program the Brazilian Open University called "Universidade Aberta do Brasil" (UAB), which sought to expand vacancies and permanence of students in public HE [45, 46]. As a result, from 2005–2014, the number of students at the federal institutions raised from 1,229,332–2,961,992 – an increment of 59.5%.

The growth of enrollments in the private sector was driven by the Prouni, started in 2004, and by the Fies instituted in 2001 [38]. Prouni grants a full scholarship upon proof of gross income monthly family allowance of up to three minimum wages ($550). Meanwhile, Fies is a student financing

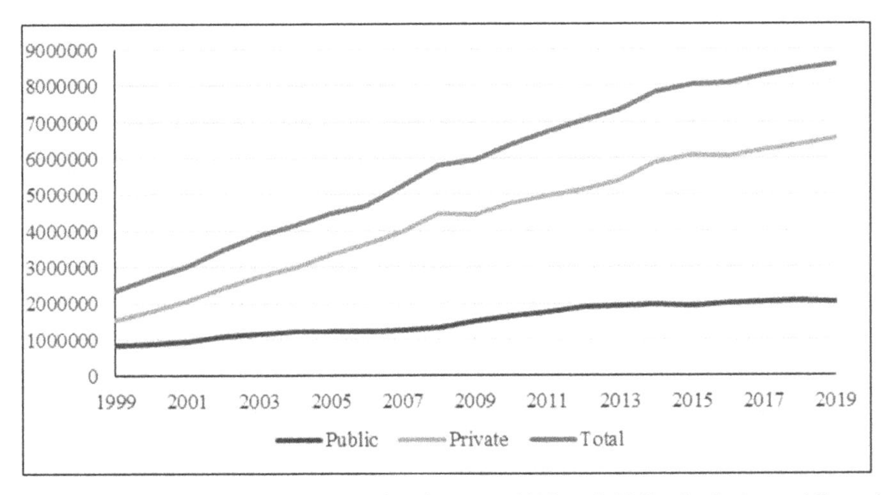

Figure 4.1 Growth of higher education between 1999 and 2019, displaying public and private sectors and on-campus and distance learning.
Source: Instituto Nacional de Estudos e Pesquisas Educacionais Anísio Teixeira (INEP), Ministry of Education of Brazil.

model that allows zero interest for students who need it most and a financial scale that varies according to the applicant's family income. One of the main trends for this expansion was DL. Figure 4.2 shows the percentual increase of distance learning higher education (DLHE) students in private institutions over the last 15 years.

The majority of HE enrollments are concentrated in the private sector in Brazil. For countries with more than 200 thousand students in HE, only Chile, Japan, and South Korea in 2014 had a higher percentage enrollment of the private sector than Brazil [47]. In the last few years, a cartelization component has been taking place in the private sector by a few private for-profit groups, with the purchase of smaller institutions and intensive use of DL. As a result, in 2018, the 10 largest for-profit private groups had 59.6% of the new enrollments (on-campus and distance learning) and 81.9% of students enrolled in distance learning courses [48].

Such growth allowed the country to get relieved of the worst rates of inclusion in HE globally compared with different countries. Table 4.1 shows the evolution of the gross enrollment rate in HE in various countries in Latin America, Asia, Europe, and North America between 2000 and 2017. This rate is given by the number of students enrolled (regardless of age) divided by young people between 18 and 23.

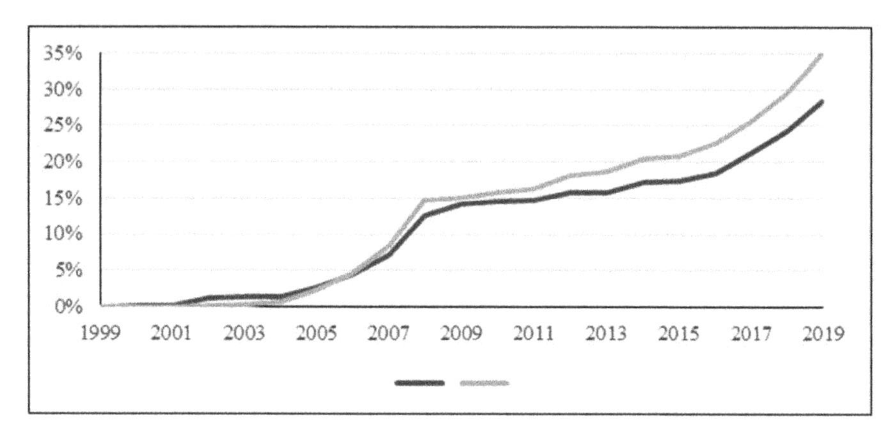

Figure 4.2 Percentual growth enrollments in distance learning between 1999 and 2019. The dark line shows the total evolvement, and the light line the private HE progression of enrollments.

Source: INEP, Ministry of Education of Brazil.

Table 4.1 Evolution of the gross enrollment rate in higher education

	2005	2008	2011	2014	2017
Brazil	**26.0**	**35.6**	**43.5**	**49.9**	**51.3**
Latin America and Caribe	31.1	39.2	43.8	47.7	51.8
Argentina	64.0	68.1	76.3	81.0	90.0
Colombia	30.7	36.1	43.0	51.4	56.4
Chile	49.4	56.7	72.3	82.8	88.5
Uruguay	45.3	50.9	..	60.7	63.1
Mexico	24.2	26.1	28.3	31.1	40.2
USA	80.6	85.0	93.9	88.6	88.2
Spain	67.2	69.7	80.0	85.4	88.9
France	53.9	52.5	55.6	61.5	65.6
Portugal	55.4	61.6	68.4	65.5	63.9
Russia	72.6	74.9	76.2	78.5	81.9
Europe	63.0	66.2	69.0	69.7	71.6
China	19.1	20.7	25.6	42.4	49.1
World	**24.3**	**27.1**	**31.4**	**35.8**	**37.9**

Source: Our World in Data (https://ourworldindata.org).

In line with the effort to expand tertiary education based on inclusion, access equity, and quality, the Ministry of Education (MEC) created, in 2005, the National Higher Education Assessment System (Sinaes). The Sinaes is responsible for assessing courses, the performance of students, and institutional profile [49]. It evaluates both on-campus and DL and uses the National

Student Performance Exam (Enade) to assess students. Accreditation and quality assessment of HEIs is mandatory by the MEC and applies equally to private and public institutions in on-campus or DL courses.

4.2.1 Social Inclusion and Access Equity

The political agenda regarding social inclusion and access equity in HE varies between countries due to economic and social priorities. The concepts of inclusion and equity that govern this plan seek to establish social justice by resolving historical inequalities in access to HE. [50] defined the common principles that govern access: the merit inherited by birth in socially favored groups, equal rights, and equity, which is commonly defined as equal opportunities. However, the massification of HE aims to reduce the principles of inherited merit, employing norms of equal rights, which resulted from the demographic, economic, political, and ideological pressures imposed by the social transformations at the end of the 20th century.

HE accessible to a more significant number of people and, more significantly, to individuals from different social backgrounds has been a challenge overcome by countless countries, particularly Brazil, where education was built on an elitist and excluding basis. Thus, the rules of equal rights have gained strength to eliminate or reduce barriers to gender, ethnic, racial, and social groups in access to HE. Bearing in mind that education alone does not solve the structural problems of inequality in countries with profound social contradictions. The university degree is a factor of self-esteem and achievement, inclusion, and social mobility. But it is recognized that inequalities are often reproduced within the university campus by areas of study and faculties [51]. Also, studies have shown that social and political disparities in access to HE, between institutions and courses, persist with solid relationships with inequalities in class, gender, race, and ethnicity [52].

Until the 2000s, Brazilian university students were mainly from the white elite and origins from the urban middle class, who attended private high school with better training for the competitive entrance exams at the best universities [53]. The class segregation pattern imposed a dualistic educational system. In the primary education system, while the middle and upper classes attend private schools, the rest of the population (about 78%) is in the public system, which remains underfunded and generally has a low quality of education. In HE, the best universities in the country are public and accessible for elites, while private colleges and paid for low-income students [33]. However, with the educational reforms introduced from the

PNE (2001–2011) and deepened with the PNE (2014–2024), this framework gained new directions in inclusion and equity in access to HE.

The number of enrollments per year in public universities has more than tripled, from 109,000 in 2006 — 393,000 in 2010, due to policies of the Federal Government based on four main actions: (1) Restructuring and Expansion Program for Federal Universities (Reuni); (2) The expansion of the offer of undergraduate courses by the Federal Technical Institutes (IFETs) and (3) The creation of the system Open University of Brazil.

The Reuni created 17 new universities and 126 new federal campuses in places that previously had no history of public universities, such as the impoverished peripheries of large cities and the semiarid region of northeastern Brazil, the Amazon region, and countless other less developed regions. Something similar occurred with the expansion of HE programs in the IFETs and the UAB, which engaged 133 public HEIs and created 777 regional distance learning centers. This expansion boosted the qualification of public services, fostered outreach programs with vulnerable communities, developed new topics in science, technology, and innovation by integrating national and global collaboration networks with local impacts [54]. The number of full-time professors at federal public universities increased by 64%.

Figure 4.3 shows the family income profile of students who graduated between 2009 and 2011 and those who graduated between 2017 and 2019, displaying a solid evolution of students from less favored social strata, with

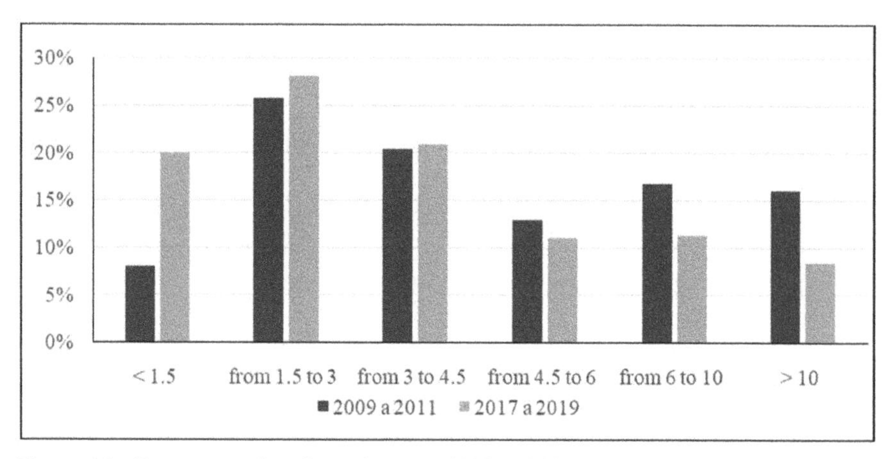

Figure 4.3 Percentage of graduates between 2009 and 2011 and between 2017 and 2019 by family income profile.

Source: INEP, Ministry of Education of Brazil.

family income less than 1.5 minimum wages ($300). Between 2009 and 2011, a total of 86,126 of the least privileged social class in Brazilian society graduated, increasing to 258,868 in the years from 2017–2019, from 8% of the total graduates to 20%.

In 2009, the Unified Selection Process (Sisu) was created to grant admission to federal public universities by the High School Quality Assessment Exam (Enem), which greatly expanded access opportunities and intranational student mobility [55]. Affirmative action policies were created, altering the profile of graduates from public universities and the environment previously almost exclusively of whites and wealthier classes [56]. Before the innovative Quota Law, several universities had initiated such policies in 2012, which established isonomic criteria for Federal Universities across the country, guaranteeing 1/2 of the enrollments for students from public high schools, including sub-criteria for ethnicity and income according to the population profile of each state. Besides, a complementary law declared the inclusion of people with disabilities in the quota system, significantly expanding inclusion and accessibility.

Thus, the student population profile at public universities came closer to the diversity of Brazilian society as a whole in terms of income, class, gender, and race. [33] revealed data from the Brazilian National Forum of Rectors of Community and Student Affairs (Fonaprace), in the period of 15 years (2003–2018), which displayed the following changes:

1. The number of students from low or medium-low-income families (less than $300 per capita per month) grew from 42.8–70.1%.
2. The number of students who come from public high school went from 36.5–60.4%, and the number of students of color (those who identify themselves as Afro-descendants or indigenous) increased from 36.2–53.5%.
3. The number of black students grew by a factor of 5.3, while the number of brown (mixed race) students increased by 3.5 in the same period.
4. The number of females grew to 48% of students, while dropped to 40% men, and 16.4% self-identified as Lesbian, Gay, Bisexual, Transgendered, and Questioning (LGBTQ) Individuals.
5. The number of female students is the majority in on-campus and distance learning modalities, aged between 19 and 21 years and 21 and 24 years.
6. The number of students with disabilities, global developmental disorders, or high skills went from 20,530 enrollments in 2009 to 48,520 in 2019.

Brazil has one of the largest public and free university systems globally, with 302 public institutions of HE (federal, state, and municipal), with 2.05 million students in 2019 [41]. The Brazilian public HE system accounts for 3/5 of postgraduate courses and 95% of the research conducted in the country, as displayed in Clarivate's report about the Brazilian scientific production between 2011 and 2016 [33]. Brazil occupies the 13th position in the production of academic research, and among the first 15 countries, it had the fifth-highest average growth rate of research between the years 2008 and 2018. Of the 25 best universities in Latin America, according to the ranking of Times Higher Education for Latin America (2019), 14 are Brazilian and almost all public. In addition to the rankings, the most important is that the turn of inclusion and equity in access in the public university system did not imply a loss of its quality standards. Also, it allowed for the first time in history to train a generation of intellectual and scientific leaders in the most diverse places and areas of knowledge, which do not come strictly from the traditionally dominant classes.

The private HE system has grown almost five times more than the public since the last decade of the 1990s, concentrating capital, attracting investment funds, and capturing public subsidies in an oligopolistic way [57]. The enormous expansion increased competitiveness, and tuition prices fell sharply until 2010. But there was evidence of a loss in the quality of education, with the transfer of private, regional, and family HEIs to the control of international investment funds and private equity. The private sector still counts with benefits from government programs, such as the Prouni and the Fies, respectively, to grant full scholarships at private institutions and lending money.

Today, most on-campus and DL enrollments are concentrated in large private educational for-profit groups, whose investment in professors is far below the public sector and small and medium-sized non-profit private institutions. The trend to reduce costs and promote universal education made distance education grow by 40% in just 4 years [58]. However, primarily low-income and working-class students have benefited from low-cost undergraduate courses and greater access to HE, including the flexibility of DL even more accessible in cost and time.

4.2.2 Professional Training in HE

The focus of HE has changed in the last 30 years, mainly in response to social and technological transformations in the labor market. The rapid increase in jobs requiring higher-order cognitive skills has created a worldwide need for more graduate-level employees [59]. Thus, the priority previously given by universities to including a small minority in research capacities has given way in many countries to providing half the population with the skills and knowledge relevant to employability. This was achieved through the rapid expansion of the HE and establishing more diverse types of HEIs.

Mass education and universal access to HE started in the 1970s in the USA and won the world over the 20th century to deal with the tremendous economic, social, and cultural differences of a growing population. For [60], the educational systems have been segmented into at least three main types: (1) elite education (0%–15% of enrollments) that shapes the mind and character of a ruling class in preparation for elite roles; (2) mass education (16%–50% of enrollments) that shapes skills and prepares the broad base of the population for the functions of the technical and economic elite; and (3) universal education (more than 50% of enrollments) that prepares the population for rapid social and technological changes.

HEIs have been structured to meet diverse educational objectives. There are undergraduate courses in multiple areas involving general or specialized training, professional training geared to the job market, academic training in research. Many offering sequential and continuing education courses to develop practical and research activities integrated with training at the undergraduate level as an instrument for preparing critical professionals and capable of permanent intellectual self-development.

There are four main types of HEIs in Brazil with a distinct purpose and social role:

(1) The federal government recognizes universities with autonomy in fulfilling requirements related to the proportion of professors with doctoral degrees employed full time, conducting research, and offering postgraduate courses at the doctoral and master's levels.
(2) University Centers, which are not yet comprehensive as universities, enjoy autonomy, expanding participation and lifelong learning with courses at all levels.
(3) Colleges carry out research and employ well-qualified teachers but generally offer only a few courses (usually, courses in medicine/health or law and studies in administration/business). They do not have the autonomy to start new programs.
(4) Colleges and technical institutes offer 3-year courses with an evident vocational characteristic.

Each of these institutions' social role and character stems from the diversification imposed by the massification of education and the high costs of investing in research. This means that not all institutions can maintain teaching, research, and continuing education activities simultaneously. Furthermore, not all institutions can maintain the same academic and quality standards. Distinct needs lead to different responses, not only in terms of products and services but also in terms of institutional arrangements or models capable of promoting such offers.

[41] statistical notes showed that in 2018, there were 2,537 HEIs, of which 299 were public and 2,238 privates, representing 88.2% of the tertiary education system (Figure 4.4). Most universities were public (53.8%), and among private universities, colleges predominate (86.2%). Of the federal HEIs, 57.3% correspond to universities, 36.4% to IFETs; 1.8% to colleges, and 4.5% are university centers. The 199 existing universities in the country are equivalent to only 7.8% of HEIs. On the other hand, 52.9% of enrollments in HE are concentrated in these universities. About 2.2% of HEIs offer 100 or more undergraduate courses, while 26.7% offer up to two undergraduate courses. Approximately 90% of undergraduate studies at universities are on-campus.

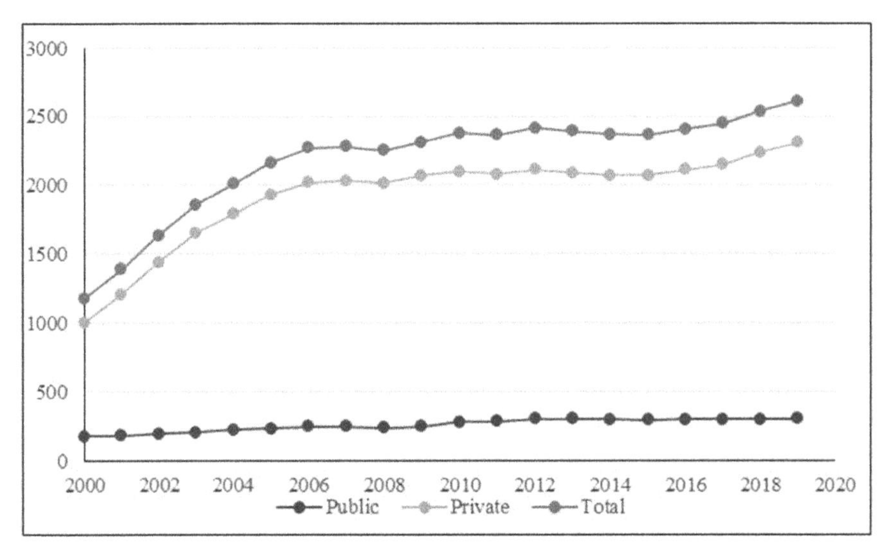

Figure 4.4 Total number of high education institutions by category.
Source: INEP, Ministry of Education of Brazil.

Besides, in 2018, almost 60% of students entering higher education chose a baccalaureate course. The baccalaureate courses continue to concentrate most students (58.0%), followed by specialized courses (20.9%) and undergraduate courses (20.5%). Between 2017 and 2018, there was an increase in students entering the bachelor's degree (3.1%). However, technical courses showed the most significant variation, 16.6%. From 2008–2018, the number of new students in technical courses registered a remarkable growth of 102.9%. The top ten undergraduate courses in the country are law, pedagogy, administration, accounting, nursing, civil engineering, psychology, physical education, medicine, and information systems.

The professional and academic technical training in all areas of knowledge has become more accessible to a broad spectrum of the Brazilian population. The country is indeed looking for professional training requirements in line with the future of work. The percentage distribution of undergraduate enrollment per area of knowledge in Brazil is comparable with the average for the countries of the Organization for Economic Cooperation and Development (OECD), except in the areas such as natural sciences, mathematics, and statistics, in which the average of OECD enrollments is higher. Around 24% of Brazilians between 25 and 34 years old hold a tertiary degree, but only 17% of the population have completed HE. Across OECD member countries, 43%, on average, hold a tertiary degree, showing that the enrollment rates of tertiary education in Brazil need to promote the further up skilling of the workforce.

4.2.3 Quality Assessment

The performance of HEIs is recognized as relevant for assessing the efficiency and effectiveness of the students' learning process. In several countries, initiatives are taken to develop models for evaluating learning programs and institutions [61]. National initiatives are essential, as they contribute as experiences in assessing learning and conceptual structures and associated methodologies. However, the variety of conceptual and methodological approaches prevents comparable metrics in an increasingly globalized environment.

Accreditation and assessment agencies worldwide have looked at comprehensive mechanisms for assessing HE quality through the performance of courses and students. For example, the OECD created the exam for newly graduated students, called Assessment of Learning Outcomes in Higher Education (AHELO), to provide data to governments, institutions, and students themselves about what they know and can do [62, 63]. In Europe, there are agency accreditation processes, as in the case of the European Association for Quality Assurance in Higher Education (ENQA). This agency, created in 2000, at the suggestion of the European Council, drew up good practice guidelines for European, national, or regional accreditation agencies [64]. To obtain the European Register of Quality Assurance Agencies (EQAR), the agencies have to go through the accreditation process carried out by ENQA, which has an expiration date. Accredited agencies guarantee the accreditation of courses and institutions in the European Area.

On the other hand, many countries have been adopting their methodologies for evaluating HE. In 2004, Brazil created the Sinaes to ensure a national assessment process for HEIs, undergraduate courses, and students' academic performance [65]. The National INEP, part of the Ministry of Education, is the federal body responsible for organizing and maintaining the educational information and statistics system and developing and coordinating Sinaes and evaluation projects in all areas, levels, and modalities of education in the country. For [66], HE systems mainly formulate their quality assurance systems according to national standards or their own needs. The public responsibility of governments to ensure quality education is based on accreditation models based on self-regulation [67].

Sinaes aims to improve HE quality, guiding the expansion of enrollments with institutional and academic, and social effectiveness. The system seeks to deepen the HEIs' social commitments and responsibilities by enhancing their public mission, promoting democratic values, respect for difference and diversity, and affirming autonomy and institutional identity. The quality analysis covers three crucial dimensions of Brazilian HE, the evaluation of undergraduate courses, the institutional assessment, and the performance evaluation of students.

[49] described the complex and dynamic character of assessing the quality of mass education, predominantly offered by the private sector. Institutions, courses, and students have access to quality as follows: (1) institutional assessment, internal and external, contemplating the global and integrated analysis of the dimensions, structures, relationships, social commitment, activities, purposes, and social responsibilities of HEIs and its courses; (2) the public character of all the procedures, data and results of the evaluation processes; (3) respect for the identity and diversity of institutions and courses; and (4) the participation of the student, teaching and technical-administrative bodies of HEIs, and civil society, through their representations. The evaluation serves as a primary reference for the processes of regulation

and supervision of HE, including the accreditation and renewal of certification of HEIs, the authorization, recognition, and renewal of recognition of undergraduate courses.

Sinaes is based on a global and integrative assessment and education concept, proposing integrating different instruments and moments of application [68]. With the evaluation result, the institution has a considerable collection of relevant information about its reality, pedagogical processes, strengths and weaknesses, and the possible ways to overcome its problems to improve performance. However, until now, the use by institutions of this information to improve performance and pedagogical course projects seems to be limited. In public universities, the obstacle seems to be the administrative discontinuity in the execution of improvement projects. In the case of for-profit private institutions, the barrier appears to increase profits by reducing costs, that is, the commercialization of education, with the consequent loss of the quality of the service delegated by the State.

The self-assessment, which is assured and valued, has not been taken seriously by most HEIs. If public universities and community, confessional, and private philanthropic universities non-profit, due to the postgraduate structure and the work regime of their faculty, have adequate conditions to carry out their internal evaluation with teachers, students, and managers. But the same cannot be said within for-profit private institutions in which hourly and part-time work of staff and professors prevails.

The latter indicates a mismatch between the evaluation process and the quality of education. In the search for increasing profit in the HE market, for-profit universities and faculties seek to meet the formal requirements of the evaluating State and consider only those minimally necessary to remain in the "market" without the risk of sanctions. This attitude distorts and limits the potential for institutional self-improvement expected by Sinaes. The consequence has been the low quality of education shown in external evaluations of the performance of their graduates at National Student Performance Exam (Enade) or the Brazilian Bar Association (OAB) Exam.

Sinaes evaluates public and private HE, both on-campus and distance learning. There is no differentiation in the treatment of institutions concerning assessment, recognition, and accreditation requirements. The university diplomas issued by the institutions do not indicate education modality with respect for students and the pedagogical model adopted. However, the diversification of public, private non-profit, and private for-profit institutions has raised important questions about the quality of education. HE in public and private institutions science-driven enjoys high national and international qualifications and an excellent reputation in society and the labor market. These universities bring together diversity and quality in education at the international level. On the other hand, the massification and rapid expansion of the private sector for profit, on-campus and distance learning, has favored the formation of oligopolies that concentrate enrollments, reduce costs and lose quality.

The assessment of student performance, conducted by the National Student Performance Exam (ENADE), is the main instrument for student assessment in Brazil [69]. ENADE is a curricular component with a mandatory census character. It was instituted by the same Law that regulated Sinaes and has a 3-year term and the ENADE cycle. The student performance is evaluated concerning the study plan provided for in the curricular guidelines of the respective undergraduate course, including their ability to adjust to the demands arising from the development of knowledge and their skills to understand topics external to the specific scope of their occupation, linked to the Brazilian reality and worldwide, and other areas of expertise. The objective is to measure the performance of enrolled students based on the skills and competencies established in the specific curricular guidelines for each course.

The Enade has a scale of 1–5, where ranges 1 and 2 are considered flawed because they are below the average in the evaluation. In 2018, 2,691 of the 6,191 courses from private institutions were evaluated by Enade and performed on grades 1 and 2, equivalent to 43.5%. Another 41.7% of the assessed courses at private institutions stayed with grades 3, while 13.4% were classified in grades 4. Only 1.5% reached the maximum grade. In 2019, Enade evaluated 8,188 courses from public and private institutions, where more than 40% of HE courses from private educational institutions (for-profit and non-profit) had an equally poor performance. Among federal universities, this rate was only 5.3%, and state public ones, 11.3%.

4.3 The Boom of Distance Learning

Brazil is one of the pioneers of DL continuing education by Edgard Roquette Pinto, who in 1923 used radio for educational programs [70]. Despite this, the offer of DL for HE started late, in 1977, with a pedagogy course at the Federal University of Mato Grosso (UFMT) and the State University (UEMT). Later, several other public HEIs throughout Brazil followed this example. In 2001, the CEDERJ consortium of public universities passed to offer open courses in the State of Rio de Janeiro [71]. After that, there was a vertiginous expansion of enrollments, mainly in the private sector, as shown in Figure 4.5.

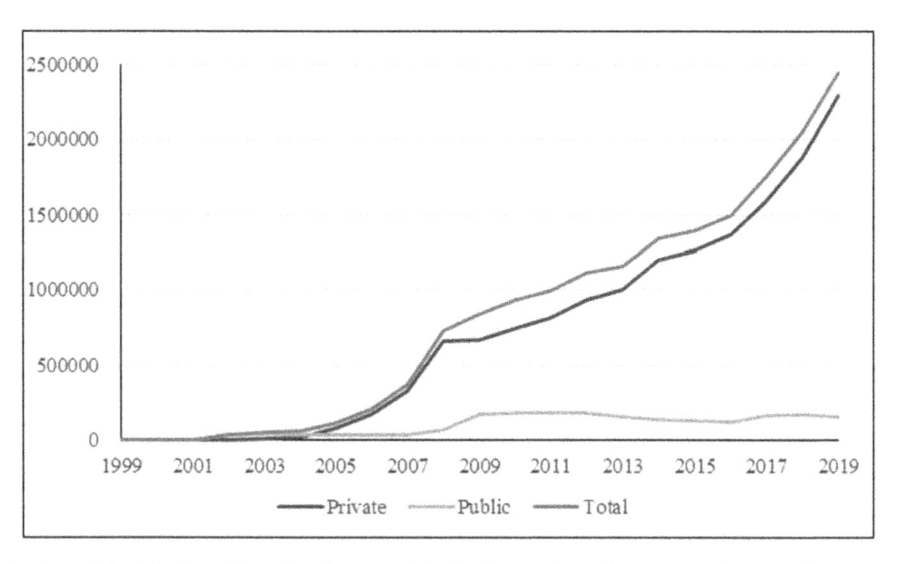

Figure 4.5 Total number of students enrolled in distance learning courses between the years 1999 and 2019.
Source: INEP, Ministry of Education of Brazil.

This expansion has one of the milestones with creating the UAB system in 2006 [46]. The UAB integrates a strategy for financing all public HEIs in the country and learning centers distributed throughout the national territory. This system, which quickly reached around 180 thousand enrollments in 2010, with potential for remarkable additional growth, has suffered from funding discontinuities, maintaining approximately the same number of enrollments since then.

DLHE enrollments have risen sharply in the last 10 years at an average rate of about 11%. The number of students enrolled in DL is equivalent to 28,.5 % of the total enrollments in HE, which means one in four. This scenario strengthens the role of DL as mass HE, particularly for older students, as shown in Figure 4.6.

On-campus enrollments have stagnated in the last 5 years, albeit with an average 10-year growth rate of 2.37% and negative rates between 2016 and 2018. However, DL enrollments showed annual growth rates of 11,.06%, reinforcing the strength of the private education in both on-campus and DL. In the last 5 years, the total enrollments in the DL modality saw an increase of 62%, while on-campus enrollments remained steady until the present. Of 2,537 HEIs in Brazil, only 14% offer DL courses. Of these, six private for-profit HEIs are responsible for more than 64% of DL enrollments.

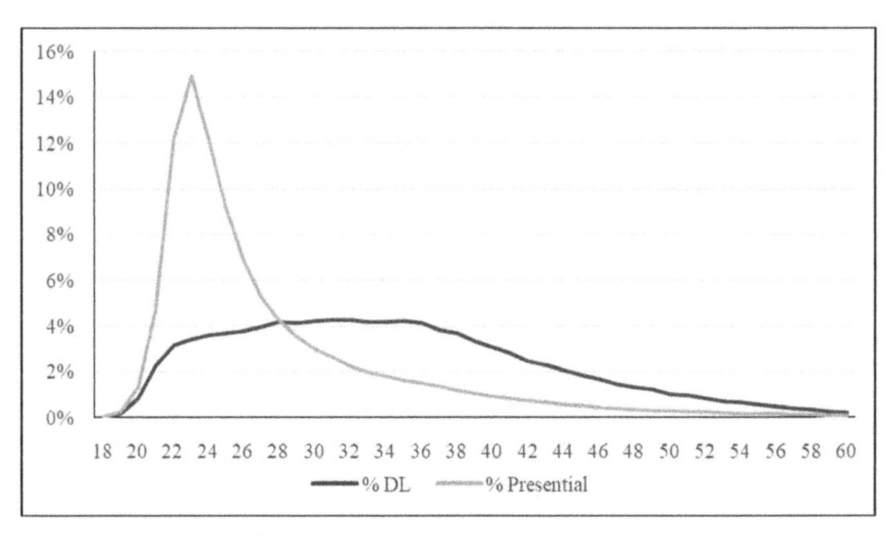

Figure 4.6 Percentage of students per age (18–60 years old) who graduated on-campus and DL between 2017 and 2019.

Source: INEP, Ministry of Education of Brazil.

In public institutions, top DL enrollments were in pedagogy (licentiate), mathematics (licentiate), public administration (bachelor), Portuguese (licentiate), and business administration (bachelor). In private institutions, the most significant number of DL enrollments was in administration (bachelor), accounting (bachelor), human resources (technologist), social services (bachelor), and physical education (licentiate). There was a high concentration of enrollments in education, the aim being to prepare future high school teachers, while the second and third choices were related to the private job market

The enrollments in administration, accounting, and human resources courses have considerably grown, confirming the trend of DL in providing mass to universal education. These courses are increasingly offered in private institutions, focusing on technical and professional qualifications and employability.

Also, in Brasil, DL has been an alternative for students who need to work and provide family assistance throughout the undergraduate course duration. The financial and time savings offered by DL figures as one of the most important social aspects of this learning modality. DL has overcome many limitations imposed by on-campus pedagogy, such as time, location, and costs, which are perceived as social gains. Besides, the DL is ubiquitous in a country with a continental dimension, involving students from rural to urban regions.

In Brazil and several other countries, investment in the physical infrastructure of education has been decreasing, and DL has already proven to be the leading provider of professional training. Thus, the PNE has awakened the interest of the population in DLHE, valuing degrees for the job market. Besides, DL has been fully adopted by companies to improve the training processes of their employees, which in the end, shows a greater openness in hiring professionals trained using this mode of learning. Data on the employability of graduates from DLHE are still scarce, but courses have increased content regarding job skills.

There is evidence of bottlenecks in courses and institutions' quality assessments, as many enrollments in some institutions may suffer from poor teaching and low-performing students [71]. The excessive concentration of the market is a problem regarding guaranteeing competitiveness and the general quality of DL. Therefore, the new regulatory framework for DL has a long way to go to reach the PNE goal in terms of market diversification since large conglomerates have acquired smaller institutions.

The growth rate of 45.50% in the total enrollments in HE between 2008 and 2018 was followed by consistent growth in DL enrollments with a total rate of 182.50% and an annual growth rate of 11.06%. In contrast with the total on-campus enrollment rate of 25.87%. Figure 4.7 compares on-campus and DL results of Enade's graduating students of the pedagogy course in two private and public institutions.

Figure 4.7 showed equivalent Enade results for on-campus and distance learning students from HEI 2 (public) and showed that Enade results for on-campus students from HEI 1 (private) are like those from HEI 2 (public). However, it showed a low performance for the students of DL course at HEI 1 (private) in 2017 with 57,688 students enrolled, indicating poor quality education based on Enade results.

The PNE's total enrollment goal of 11 million students might well be reached, most likely in 2026 if DL enrollments maintain an annual growth rate of 11.06%. Moreover, the DL enrollments may exceed on-campus enrollments from 2026 onwards. This indicates the remarkable contribution of DLHE in achieving the PNE goal and creates pressure on the educational system looking for quality assessment. There is a strong tendency for DL enrollments to grow in the future but in the face of high-quality issues.

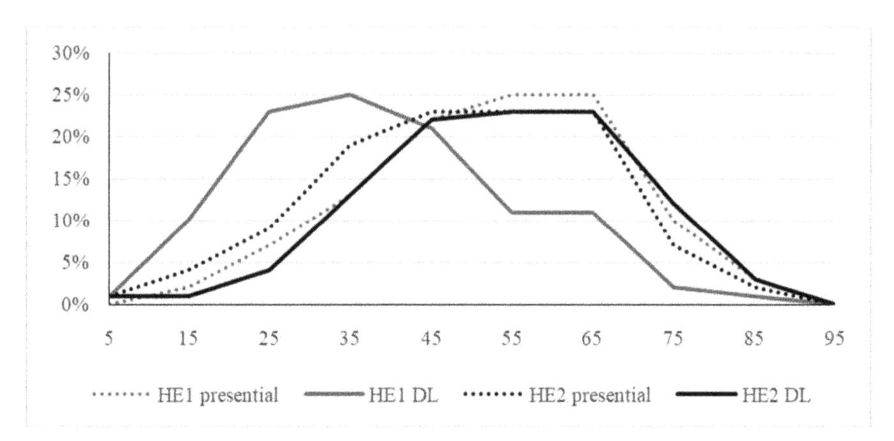

Figure 4.7 Enade of graduating students of pedagogy course in two HEIs, presential and DL modalities. Each learning modality represents 25% of the total, and Enade grades vary from 5 to 95.
Source: INEP, Ministry of Education of Brazil.

4.4 Looking into The Future

The current picture of HE in Brazil results from the guidance of the PNE (2001–2011) and PNE (2014–2024) to expand enrollments in post-secondary education, promoting greater social inclusion and equity in access concerning gender, ethnicity, and socio-economic status. It is in line with the principles of SDG 4, as it aims to train young people and adults in technical and professional skills to achieve decent work and employment, including in their curricula concepts of sustainable development, such as climate change, human rights, and world peace.

Brazil has kept an educational policy in full swing with reforms necessary to place the country in the world of work in the 21st century. These reforms, initiated in the late 1990s, were aimed at mass education and the universalization of HE, therefore imbued by reducing inequalities in access and professional training.

As [26] attested, the mass HE cannot end inequalities but can serve as an engine of development and progress. More students at universities and colleges allow social mobility, improving the quality of life for families of social groups previously excluded from access to HE. Inequalities can only be resolved with policies in other sectors such as health, employment, income distribution, and social justice.

As of 2026, Brazil should reach 11 million students between 18 and 24 years of age enrolled in HE, which partially meets the PNE target

(2014–2024). This goal indicates significant advances in training a large part of the young population at a global level compatible with sustainable development. In two decades, HE grew 256%, from 2.37 million enrollments in 1999 — 8.45 million in 2018.

Between 1999 and 2019, the public sector grew almost 40% in enrollments, from 832,022 in 1999–2,056,460 in 2019, with 1.898.803 on-campus and 157.657 in DL. Thanks to Reuni and the plan of assistance, such as the National Student Assistance Plan (Pnaes) and the Law of Ethnic Quotas, the profile of the students gained the nuance of the Brazilian diversity in the best universities in the country. Public and free institutions doubled the number of students, maintained a high level of teaching, an international group of scientific research, and remained as radiating hubs of services for the community and projects of social and entrepreneurial basis. Public universities also represent the forefront of the country in the international mobility of students and researchers.

The private sector has grown faster than the public sector since 1999, reaching a total of 6,548,066 enrollments in 2019, with 4,255,458 on-campus and 1,898,803 DL courses. While the public sector has maintained approximately the supply profile in the last 10 years, the private sector has changed a lot, mainly due to increased enrollments in a few educational for-profit groups and higher percentage participation of students in DL. In 2018, the 10 largest private for-profit educational groups had 59.6% of new enrollments (on-campus and DL) and 81.9% of students enrolled in DL courses. This indicates a strong tendency for the private sector for-profit to occupy all market share of DL.

It is a fact that Brazil has taken a significant turn to enable broad access to HE in a diversified and inclusive way. For many, this turn had a form of credentialism, but not inclusion and equity, since inequalities remain, and the quality of education is not guaranteed institutionally for all. However, there is evidence that massification promoted advances in transforming a generation of young Brazilians whose parents probably did not even finish high school. Above all, today, young people seek university education and greater knowledge, employability, and wages. There is greater confidence that they may have access to a university degree and professional training at the tertiary level.

Undeniably, there is a need for a revision in the expansion of Brazilian HE, which exposes weaknesses, scars, and risks after two decades. It is essential to project the future and build an even more inclusive quality HE for lifelong learning. The challenges for the future are plural and complex, so it is impossible to cover all topics here. However, it seems necessary to highlight those most sensitive to the objectives of sustainable development: quality of education, pedagogical model and curricular content, and investment in teaching and research.

The quality assessment promoted by Sinaes contains advances on institutional self-assessment, visits by specialists in loco, and the development of institutional and pedagogical plans. However, this system has not adequately assessed so differentiated institutions operating on-campus and DL. Public universities actively develop teaching and research, having a full-time doctoral staff. These universities have the best results in qualifying students in the Enade and national and international recognition. On the contrary, private for-profit universities follow evaluation rules at the lower limit of the quality range. Besides, there is a broad spectrum of private, community, denominational and philanthropic institutions, with variable academic performance between public universities and private for-profit universities. Many private non-profit institutions have course performance comparable to that of public universities. However, the large concentration of enrollments in only ten private for-profit universities with low academic performance, both on-campus and DL courses. If this trend prevails in the future, it may compromise the quality of HE in Brazil as a whole [48].

China has made significant contributions to the transformation of HE from an elite system to a widespread system through DL [21]. India had a gross DL enrollment ratio of 26% in 2017, with over 35 million HE students, but the country faces national leadership challenges with various open universities [72]. In South Africa, DL should increase as more traditional campus-based institutions provide DL opportunities [73]. DL constitutes an essential part of development in Turkey as the central pillar in providing HE for the masses and educational practice for lifelong learning [74]. DL education has been the primary vehicle in many countries to transition from an elite to a mass system of HE [75]. Moreover, the DL modality requires its process of quality assessment of education in Brazil and the world, as it is a unique pedagogical model with a very differentiated academic infrastructure, involving qualified not only professors but also many specialized tutors by students and distance learning centers, for on-campus practical research and classroom activities. This is necessary for validating the DLHE under the scope of SDG 4 and the PNE (2014–2014).

The sole role of the State and the internal commissions of the Ministry of Education (MEC) in the process of evaluating undergraduate education has not been entirely capable of facing the oligopoly formed by educational groups, much less sanctioning loss of quality and serving as guidance the improvement of quality. Evaluating the quality of institutions by peers is an alternative very successful experience in Brazil for postgraduate courses at the masters and doctoral level. This mechanism has given the Coordination for the Improvement of Higher Education Personnel (CAPES) great international credibility in training young masters and doctors and has contributed to making Brazilian scientific production competitive on the global stage. The peer-review model maintains public transparency about previously established assessment criteria and also a dynamic to stimulate competitiveness by quality. It is a way of guaranteeing institutional autonomy and contributing to the State in a regulatory role.

The objective of Enade is to measure student's performance concerning the syllabus contents foreseen in the curricular guidelines of courses and the comprehension of themes related to the Brazilian and international reality. The exam allows institutions to use their results as an ingredient of an internal evaluation process to improve the course's pedagogical project. In addition, the exam is the most important indicator that society uses to assess the quality of the classes, giving rise to the Preliminary Course Concept (CPC), which guides the "in loco" visits made by Sinaes. However, alongside these aspects, it is necessary to consider that a test with 40 items is not enough to assess sufficient knowledge to exercise a profession. In addition, due to its 3-year application, it does not allow one to follow classes at each stage of learning. Other difficulties are added, such as analyzing the differences between first-year students and graduates from a single exam. An exciting proposal would be to include, in the questionnaires, other dimensions of research that effectively raise the contributions of HE to students' performance. Pretesting the exam items and questionnaires, from a technical point of view, would favor the quality of the instruments and the comparability of the results. There is a need for additional assessment forms due to the difficulty of a single standardized test that does not fully assess students' learning and skills.

Furthermore, Sinaes were built when the country did not have a significant offer of DLHE courses. The quality assessment was then mainly focused on on-campus HE. The CPC is one of the leading indicators based on faculty qualification. It is based on the percentage of doctors and the percentage of full-time professors, regardless of the number of professors per student. However, there are cases of DL courses with more than 1.3 thousand students per professor. Thus, the low performance of students in the Enade is not compatible with the qualification of professors regarding the CPC. Another indicator that has shortcomings for DL with 35% of the weight in the CPC is the difference between the knowledge acquired by students at the entrance and graduation, measured by Enade [76].

There is also the challenge of monitoring students graduating from HE and their performance in professional life. Few Brazilian institutions maintain an alumni association to guide internal career programs and changes in the curriculum of the courses. There is no feedback from the labor market concerning graduates, and the curricula of undergraduate courses seem frozen in time. Education should provide students with the skills they need to succeed, and the emotional intelligence required to work in groups and diversity. Creativity, critical thinking, communication, and collaboration are concepts that must appear in all curricula, along with language and mathematics. Investigative problem-solving skills are essential for the development of advanced knowledge and confidence. Current education needs to provide students with transferable skills that will empower them in a rapidly changing world. [78] defend the end of our old-fashioned industrial education system and propose an organic approach that uses today's unprecedented technological and professional resources to engage all students, develop their love of learning and enable them to face the real challenges of the future.

The new generation of students is fully engaged in the online environment with access to mobile devices, making learning much less centralized. Access to information made the teaching structure more democratic and less dependent on the school's physical environment. Open distance learning is an example of the educational revolution made possible by new technologies. Traditional classroom classes are often considered monotonous and unproductive. Therefore, it is a consensus that educational institutions that do not follow this educational revolution will be at serious risk of being stuck in the past for following a teaching methodology that no longer meets the needs of students.

Universities represent a source of knowledge and experimentation, producing and disseminating knowledge as a basis for sustainable development [79]. The performance of universities should focus on learning and teaching, providing students with knowledge, skills, and motivation to understand and address the SDGs within a context of education for sustainable development. Universities must be compromised with inclusive education for all, including social and racial quotas. Also, provide efforts in research for raising the knowledge, solutions, and technologies that enable the visualization of possible paths for the implementation of the SDGs. Universities have a leading role in public engagement and participation in decision-making concerning the SDGs. The Times Higher Education ranking in 2020 listed, through self-declarations, the universities best classified in terms of performance with the implementation of the SDGs. The Brazilian leadership is present in all the SDGs in 12 public universities. This perspective should influence other universities in the country and the future evaluation processes.

Above all, the guarantee of public investment in education and research is essential to meet the expansion of enrollments and HE's quality in Brazil. The PNE (2014–2024) predicted an increase in the GDP for education up to 10%, which will be postponed due to the economic downturn and liberal government policy. This increase is as substantial as achieving greater efficiency and effectiveness in education at all levels, especially quality. Besides, the strengthening of for-profit private education as oligopolies may bring down competitiveness and low-quality education standards.

Most countries have successfully expanded the education system. Massification provided a differentiated academic system with various institutions with different purposes and levels of quality, which is compromised with the SGD 4. The research university still figures at the top of educational systems, but it is no longer the only model for post-secondary education. Wealthier, higher-income countries educate more than 30 % of young people in the relevant age group in post-secondary education. Many developing countries have doubled access, such as Brazil, but there is a long way to fulfill all requirements of sustainable development. The wave of expansion of education worldwide is the foundation to build knowledge and skills for a new society, which can live a more humane and sustainable future for the planet.

References

[1] Keeble BR. The Brundtland report: 'Our common future.' Med War. Jan. 22; vol. 4, no. 1, pp. 17–25, 1988.
[2] Visser W, Brundtland GH. Our Common Future ('The Brundtland Report'): World Commission on Environment and Development. In: The Top 50 Sustainability Books. Greenleaf Publishing Limited; p. 52–5, 2013.
[3] United Nations. United Nations Millennium Declaration - A/RES/55/2. Gen Assem. (18 Sep.), 2000.
[4] Unesco. The Dakar Framework for Action. Unesco; (Apr.), pp. 26–8, 2000.

[5] McCowan T. Three dimensions of equity of access to higher education. Comp A J Comp Int Educ. Jul. 3; vol. 46, no. 4, pp. 645–65, 2016.
[6] Wu Y-CJ, Shen J-P. Higher education for sustainable development: a systematic review. Int J Sustain High Educ. Sep. 5; vol. 17, no. 5, pp. 633–51, 2016.
[7] Mense EG, Lemoine PA, Garretson CJ, Richardson MD. The Development of Global Higher Education in a World of Transformation. J Educ Dev. Sep. 20; vol. 2, no. 3, pp. 47, 2018.
[8] Noui R. Higher education between massification and quality. High Educ Eval Dev. Nov. 18; vol. 14, no. 2, pp. 93–103, 2020.
[9] United Nations. The 2030 Agenda for Sustainable Development. New York, USA, 2015.
[10] Owens TL. Higher education in the sustainable development goals framework. Eur J Educ. Dec., vol. 52, no. 4, pp. 414–20, 2017.
[11] Albareda-Tiana S, Vidal-Raméntol S, Fernández-Morilla M. Implementing the sustainable development goals at University level. Int J Sustain High Educ. Mar. 5; vol. 19, no. 3, pp. 473–97, 2018.
[12] Trow M. Reflections on the Transition from Elite to Mass to Universal Access: Forms and Phases of Higher Education in Modern Societies since WWII. In: International Handbook of Higher Education. Dordrecht: Springer Netherlands; p. 243–80, 2008..
[13] Oketch M. Financing higher education in sub-Saharan Africa: some reflections and implications for sustainable development. High Educ. Oct. 18; vol. 72, no. 4, pp. 525–39, 2016.
[14] Marginson S. The worldwide trend to high participation higher education: dynamics of social stratification in inclusive systems. High Educ. Oct. 2; vol. 72, no. 4, pp. 413–34, 2016.
[15] Levy DC. The Decline of Private Higher Education. High Educ Policy. Mar. 6; vol. 26, no. 1, pp. 25–42, 2013.
[16] Yang L, McCall B. World education finance policies and higher education access: A statistical analysis of World Development Indicators for 86 countries. Int J Educ Dev. Mar.; vol. 35:25–36, 2014.
[17] Buckner E. The Worldwide Growth of Private Higher Education: Cross-national Patterns of Higher Education Institution Foundings by Sector. Sociol Educ. Oct. 25; vol. 90, no. 4, pp. 296–314, 2017.
[18] Hauschildt K, Gwosć C, Netz N, Mishra S. Social and Economic Conditions of Student Life in Europe. Hauschildt K, editor. Bielefeld: W. Bertelsmann Verlag GmbH & Co. KG; 225 pp, 2015.

[19] Qayyum A, Zawacki-Richter O. The State of Open and Distance Education. In: Zawacki-Richter O, Qayyum A, editors. Open and Distance Education in Asia, Africa and the Middle East National Perspectives in a Digital Age. SpringerBriefs in Education Open and Distance Education: Springer Singapore; p. 125–40, 2019.
[20] Tekneci PD. Evolution of the Turkish higher education system in the last decade. J High Educ Sci. vol. 6, no. 3, pp. 277, 2016.
[21] Li W, Chen N. China. In: Zawacki-Richter O, Qayyum A, editors. Open and Distance Education in Asia, Africa and the Middle East National Perspectives in a Digital Age. SpringerBr. Singapore: Springer Singapore; p. 7–23, 2019.
[22] Schendel R, McCowan T. Expanding higher education systems in low- and middle-income countries: the challenges of equity and quality. High Educ. Oct. 18; vol. 72, no. 4, pp. 407–11, 2016.
[23] McCowan T. Universities and the post-2015 development agenda: an analytical framework. High Educ. Oct. 18; vol. 72, no. 4, pp. 505–23, 2016
[24] Brennan J, Naidoo R. Higher education and the achievement (and/or prevention) of equity and social justice. High Educ. Sep. 18; vol. 56, no. 3, pp. 287–302, 2008.
[25] Tomlinson M, Watermeyer R. When masses meet markets: credentialism and commodification in twenty-first century Higher Education. Discourse Stud Cult Polit Educ. Sep. 6; vol. 1–15, 2020.
[26] Altbach PG, Reisberg L, Rumbley LE. Tracking a Global Academic Revolution. Chang Mag High Learn. Feb. 26; vol. 42, no. 2, pp. 30–9, 2010.
[27] Salmi J, D'Addio A. Policies for achieving inclusion in higher education. Policy Rev High Educ. Oct. 20; vol. 1–26, 2020
[28] Collste D, Pedercini M, Cornell SE. Policy coherence to achieve the SDGs: using integrated simulation models to assess effective policies. Sustain Sci. Nov. 26; vol. 12, no. 6, pp. 921–31, 2017.
[29] Ferguson T, Iliško D, Roofe C, Hill S. Inclusivity, Equity and Lifelong Learning For All. In: Leal Filho W, Mifsud M, editors. SDG4 – Quality Education. First. Bingley, UK: Emerald Publishing Limited; 122 pp, 2019.
[30] Beisiegel C de R. O plano nacional de educação. Cad Pesqui. Mar.; no. 106, pp. 217–31, 1999.
[31] Bucci MPD, Gomes FAD. A piece of legislation for the guidance of public education policies in Brazil: the National Education Plan 2014–2024. Theory Pract Legis. Sep. 2; vol. 5, no. 3, pp. 277–301, 2017.

[32] Ristoff D. O novo perfil do campus brasileiro: uma análise do perfil socioeconômico do estudante de graduação. Avaliação Rev da Avaliação da Educ Super. Nov.; vol. 19, no. 3, pp. 723–47, 2014.
[33] Arantes PF. Higher education in dark times: from the democratic renewal of Brazilian universities to its current wreck. Policy Rev High Educ. Jan. 22; vol. 1–27, 2021.
[34] Borsato FP, Alves J de M. Student Assistance in Higher Education in Brazil. Procedia - Soc Behav Sci. Feb. ; vol. 174:1542–9, 2015.
[35] Kirakosyan L. Affirmative action quotas in Brazilian higher education. J Multicult Educ. Jun. 3; vol. 8, no. 2, pp. 137–44, 2014
[36] Schwartzman LF, Paiva AR. Not just racial quotas: affirmative action in Brazilian higher education ten years later. Br J Sociol Educ. May 18; vol. 37, no. 4, pp. 548–66, 2016.
[37] Felicetti VL, Morosini MC, Somers P. Affirmative Action in the Quality of Higher Education: The Voices of Graduates of the University for All Program. Policy Futur Educ. Aug.; vol. 11, no. 4, pp. 401–13, 2013.
[38] Chaves VLJ, Amaral NC. Política de Expansão da Educação Superior no Brasil - O Prouni e o Fies como financiadores do setor privado. Educ em Rev. Dec.; vol. 32, no. 4, pp. 49–72, 2016.
[39] Santos CDA. Educação Superior a Distância no Brasil: democratização da oferta ou expansão do mercado. Rev Bras Política e Adm da Educ - Periódico científico Ed pela ANPAE. Apr. 30; vol. 34, no. 1, pp. 167, 2018.
[40] Giolo J. A educação a distância e a formção de professores. Educ e Soc. vol. 29, no. 105, pp. 1211–34, 2008.
[41] INEP. Censo da Educação Superior: Notas estatísticas 2018. Instituto Nacional de Educação e Pesquisas Anísio Teixeira - Ministério da Educação - Diretoria de estatísticas educacionais, 2019.
[42] Aguiar MA da S. Avaliação do Plano Nacional de Educação 2001–2009: questões para reflexão. Educ Soc. Sep.; vol. 31, no. 112, pp. 707–27, 2010.
[43] INEP. Plano Nacional de Educação PNE 2014-2024 Linha de Base. Brasilia, DF, Brasil, 2014.
[44] Schwartzman S. Education in South America. 1st ed. Schwartzman S, editor. Education Around the World. Bloomsbury Publishing Plc; 480 pp, 2015.
[45] Favato MN, Ruiz MJF. REUNI: política para a democratização da educação superior? Rev Eletrônica Educ. 448–63, 2018.

[46] Costa CJ da, Pimentel NM. O sistema Universidade Aberta do Brasil na consolidação da oferta de cursos superiores a distância no Brasil. ETD - Educ Temática Digit. Oct. 7; vol. 10, no. 2, pp. 71, 2009.
[47] Levy DC. Global private higher education: an empirical profile of its size and geographical shape. High Educ. Oct. 30; vol. 76, no. 4, pp. 701–15, 2018.
[48] Bielschowsky CE. Tendncias de precarizao do ensino superior privado no Brasil. Rev Bras Poltica e Adm da Educ - Peridico cientfico Ed pela ANPAE. May 11; vol. 36, no. 1, 2020.
[49] Pereira CA, Araujo JFFE, de Lourdes Machado-Taylor M. The Brazilian higher education evaluation model: "SINAES" sui generis? Int J Educ Dev. 61 (Jul. 2017), pp. 5–15, 2018.
[50] Clancy P, Goastellec G. Exploring Access and Equity in Higher Education: Policy and Performance in a Comparative Perspective. High Educ Q. Apr.; vol. 61, no. 2, pp. 136–54, 2007.
[51] Bilecen B, Van Mol C. Introduction: international academic mobility and inequalities. J Ethn Migr Stud. Jun. 11; vol. 43, no. 8, pp. 1241–55, 2017.
[52] Gisi. ML. A Educação Superior no Brasil e o caráter de desigualdade do acesso e da permanência. Rev Diálogo Educ. Jul. 17; vol. 6, no. 17, pp. 97, 2006.
[53] Childs P, Stromquist NP. Academic and diversity consequences of affirmative action in Brazil. Comp A J Comp Int Educ. Sep. 3; vol. 45, no. 5, pp. 792–813, 2015.
[54] Niquito TW, Ribeiro FG, Portugal MS. Impacto Da Criação Das Novas Universidades Federais Sobre As Economias Locais. Planej e Políticas Públicas. , no. 51, pp. 367–94, 2018.
[55] Andriola WB. Doze motivos favoráveis à adoção do Exame Nacional do Ensino Médio (ENEM) pelas Instituições Federais de Ensino Superior (IFES). Ens Avaliação e Políticas Públicas em Educ. Mar.; vol. 19, no. 70, pp. 107–25, 2011.
[56] Passos JC dos. Relações raciais, cultura acadêmica e tensionamentos após ações afirmativas. Educ em Rev. Jun.; vol. 31, no. 2, pp. 155–82, 2015.
[57] Chaves VLJ. Expansão da privatização/mercantilização do ensino superior Brasileiro: a formação dos oligopólios. Educ Soc. Jun.; vol. 31, no. 111, pp. 481–500, 2010.

[58] Giolo J. Educação a Distância no Brasil: a expansão vertiginosa. Rev Bras Política e Adm da Educ - Periódico científico Ed pela ANPAE. Apr. 30; vol. 34, no. 1, pp. 73, 2018.

[59] Mello SL de M, Ludolf NVE, Quelhas OLG, Meiriño MJ. Innovation in the digital era: new labor market and educational changes. Ens Avaliação e Políticas Públicas em Educ. Mar. 1; vol. 28, no. 106, pp. 66–87, 2020.

[60] Trow M. Problems in the Transition from Elite to Mass Higher Education. Int Rev Educ. 18:61–82, 1973.

[61] van Damme D. Trends and models in international quality assurance in higher education in relation to trade in education. High Educ Manag Policy. Dec. 17; vol. 14, no. 3, pp. 93–136, 2002.

[62] Morgan C, Shahjahan RA. The legitimation of OECD's global educational governance: examining PISA and AHELO test production. Comp Educ. Apr. 3; vol. 50, no. 2, pp. 192–205, 2014.

[63] Richardson S, Coates H. Essential foundations for establishing an equivalence in cross-national higher education assessment. High Educ. Dec. 19; vol. 68, no. 6, pp. 825–36, 2014.

[64] Blomqvist C, Donohoe T, Kelo M, Linde KJ, Llavori R, Maguire B, et al. Quality Assurance and Qualifications Frameworks: Exchanging Good Practice. ENQA Workshop Report 21. ENQA (European Association for Quality Assurance in Higher Education). pp. 1–44 p, 2012.

[65] Pedrosa RHL, Simões TP, Carneiro AM, Andrade CY, Sampaio H, Knobel M. Access to higher education in Brazil. Widening Particip Lifelong Learn. May 1; vol. 16, no. 1, pp. 5–33, 2014.

[66] Alzafari K, Ursin J. Implementation of quality assurance standards in European higher education: does context matter? Qual High Educ. 25, no. 1, pp. 58–75, 2019.

[67] Stensaker B, Langfeldt L, Harvey L, Huisman J, Westerheijden D. An in-depth study on the impact of external quality assurance. Assess Eval High Educ. Jul.; vol. 36, no. 4, pp. 465–78, 2011.

[68] Dias Sobrinho J. Avaliação e transformações da educação superior brasileira (1995—2009): do provão ao Sinaes. Avaliação Rev da Avaliação da Educ Super. 15, no. 1, pp. 195–224, 2010.

[69] Brito MRF de. O SINAES e o ENADE: da concepção à implantação. Avaliação Rev da Avaliação da Educ Super. Nov.; vol. 13, no. 3, pp. 841–850, 2008.

[70] Almeida DS de, Marques PF. MOOCSă: uma análise das experiências pioneiras no Brasil e Portugal - constatações e limitações. In: Congresso

Internacional ABED de Educação a Distância, Bento Gonçalves, RS, Brasil, 2015.
[71] Bielschowsky CE. Revista Científica em Educa*o* a Distância. Ead Em Foco. 7, no. 2, pp. 8–27, 2010.
[72] Oliveira ÉT De, Piconez SCB. Avaliação da educação superior nas modalidades presencial e a distância: análises com base no Conceito Preliminar de Cursos (CPC). Avaliação Rev da Avaliação da Educ Super. Dec.; vol. 22, no. 3, pp. 833–51, 2017.
[73] Sharma RC. India - Commentary. In: Zawacki-Richter O, Qayyum A, editors. Open and Distance Education in Asia, Africa and the Middle East National Perspectives in a Digital Age. SpringerBr. Singapore: Springer Singapore; p. 43–7, 2019.
[74] Prinsloo P. South Africa. In: Zawacki-Richter O, Qayyum A, editors. Open and Distance Education in Asia, Africa and the Middle East National Perspectives in a Digital Age. SpringerBr. Singapore; p. 67–83, 2019.
[75] Selvi K. Right of Education and Distance Learning. Eurasian J Educ Res. 22:201–1, 2006.
[76] Cooperman L. From elite to mass to universal higher education: from distance to open education. RIED Rev Iberoam Educ a Distancia. Jan. 28; vol. 17, no. 1, 2014.
[77] Bielschowsky CE. Análise dos Resultados do Exame Nacional de Desempenho de Estudantes (Enade) para Educação a Distância do ciclo 2015 a 2017. EaD em Foco. Oct. 31; vol. 8, no. 1, 2018.
[78] Robinson K, Aronica L. Creative Schools: Revolutionizing Education from the Ground Up. Reprint. Penguin Books, Limited; 320 pp. (Penguin books), 2016.
[79] Sedlacek S. The role of universities in fostering sustainable development at the regional level. J Clean Prod. Jun.; vol. 48:74–84, 2013.

5

Incorporating SDG 11 in Higher Education Teaching – The Relevance of Mobility on Sustainable Cities and Communities

Margarida C. Coelho

University of Aveiro – Dept. Mechanical Engineering/
Centre for Mechanical Technology and Automation, Portugal
ORCID ID: 0000-0003-3312-191X
E-mail: margarida.coelho@ua.pt
Corresponding Author

Abstract

The mobility sector is currently facing a series of new paradigms (from driverless, electric, shared, connected vehicles, to micro-mobility), combined with the challenges of efficient, inclusive, safe and sustainable mobility. There are many challenges arising in the forthcoming years: according to the European Commission Sustainable and Smart Mobility Strategy, the objectives by 2030 are that at least 30 million zero-emission cars will be in operation on European roads, 100 European cities will be climate neutral and automated mobility will be deployed at large scale. By 2050 the intention is that nearly all cars, vans, buses and new heavy-duty vehicles will be 'zero-emissions'. However, the dependency on conventional fuels is still a reality for road transport. About 73% of greenhouse gases emissions from the transport sector are related to road transport in Europe. Also, around 400,000 deaths in Europe per year are attributable to air pollution.

This book chapter addresses the future challenges inherent in the incorporation of the mobility for more sustainable and resilient cities on higher education, responding to the United Nations Sustainable Development Goal 11 (Sustainable Cities and Communities), namely regarding the reduction of negative impacts, such as greenhouse gas emissions, emissions of nocive pollutants, traffic congestion, road safety and noise. The main objectives of this book chapter will be threefold: 1) to describe how to incorporate smart and sustainable mobility-related topics in higher education projects; 2) to tie the teaching/learning process with the research developed by the author, involving the implementation of smart mobility applications and innovative mobility approaches to pave the way for sustainable cities and communities.

Keywords: Mobility, Transport, Higher Education Institutions, Teaching, Research.

5.1 Introduction

Transport systems and mobility are of utmost importance for the challenge of competitiveness and the development of the economy, as well as the need to ensure mobility and accessibility to people and goods, efficiently and adequately, promoting social inclusion. In addition to the relevance of the topic and as society recognizes the growing importance of optimizing transport systems, it is necessary to know how to identify and evaluate innovations that arise in the mobility of people and goods. In this way, and taking into account the higher education institutions (HEI) commitment to be at the service of science and society, the scientific adequacy of the thematic field 'Smart Mobility and Transport' is demonstrated to the interests and needs of many HEI pedagogic and scientific strategies.

Thus, why a book chapter entitled 'Incorporating SDG 11 in Higher Education Teaching – The relevance of mobility on sustainable cities and communities'? The three main facts that motivated the author to propose this book chapter are listed below:

1. Relevance of the transport sector in the economy: the importance of transportation and related mobility for a whole economy (macroeconomic level, in which transportation is linked to a level of output, employment, and income within a national economy) accounts for 6%–12% of the GDP in many developed economies [1]. Also, logistics costs can account for between 6% and 25% of the GDP [1]. Overall, all transportation assets (including infrastructures and vehicles) can account for about half the GDP of an advanced economy. From a microeconomic level perspective (the importance of transportation for specific parts of the economy), transportation is linked to producer, consumer and

distribution costs and, on average, transportation accounts for between 10% and 15% of household expenditures [1].

2. Relevance of fuel consumption in the transport sector (when compared to other sectors of activity), as well as the instability in the respective prices. It is essential to reduce its energy intensity to promote competitiveness. It is a sector that depends heavily on oil and petroleum products. Reducing this dependency is a necessity and a technological challenge.
3. Need to reduce greenhouse gas (GHG) emissions and local pollutants (pollutants that cause a direct and negative impact on human health), given the importance of the transport sector on emissions [2–4].
4. Relevance of other impacts on the transport system, such as traffic congestion and road safety, highlights the complexity of the sector.

Thus, there is plenty of interest to include mobility-related topics in the syllabus of higher education courses and degrees, namely to:

1. Tackle a permanent development of a dynamic teaching pedagogy on HEI, with updated contents and learning materials, making them available to students;
2. Contribute to the development of students' critical, inventive and creative spirit, supporting them towards an autonomous and rigorous posture, as well as stimulating them towards cultural, scientific, professional and human training;
3. Develop the students' cultural and scientific knowledge;
4. Articulate connections between HEI and their local, regional and national stakeholders in the field of 'Sustainable Mobility', in order to act as catalysts for regional development and innovation for the communities.

Additionally, from the point of view of the author, the incorporation of the UN Sustainable Development Goals should encourage the development of reasoning skills and cognitive reflection by the student, in a constructive way. Thus, it is vital that the author contributes to develop skills in students, namely in the way in which work is investigated and developed in the field of mobility systems and technologies. Thus, the ultimate goal of this approach in the teaching-learning process will be an effective 'learn how to do' practice that encourages critical reflection and discussion of the different topics covered in the curricular unit. For the acquisition of these skills, explanations about a given domain will have to be substantiated, to overcome a given mobility challenge.

5.2 Incorporating Mobility in the Teaching and Research Activities of an HEI

The main objective of incorporating mobility-related topics in HEI is to empower students with the knowledge and skills to understand the existing mobility challenges and some of the smart and sustainable solutions that can be implemented. It is intended that students, during this phase, should be able:

- To understand the present situation of mobility and the transportation sector;
- To identify sustainable mobility solutions, measures and new paradigms;
- To identify technologies that can be implemented to tackle smart and sustainable mobility.

The themes of energy and transport have a very strong global dimension and converge in a societal challenge of the European Commission in Horizon 2020: 'Smart, Ecological and Integrated Transport'. The intention of this section is to share learning experiences carried out in conjunction with research carried out within the scope of this societal challenge. Thus, this section will focus on the results of two initiatives, with the common objective of achieving a coherent posture in the set of teaching-research activities.

The first initiative concerns the course 'Energy, Mobility and Transport', which is part of the curricular plan of the several Master degrees at the University of Aveiro (such as Master on Smart Mobility, Mechanical Engineering, Sustainable Energy Systems and Environmental Engineering). The course topics are interdisciplinary and contribute to the integration of results and international research practices in the students' curriculum. Thus, the final objective of this approach is to encourage critical reflection and the scientific spirit in the different topics covered, through the articulation between the thematic contents of the course and the scientific competences of the Professor. Above all, it is intended that all students understand the subjects without misplacing the scientific accuracy of the contents.

The second initiative brings the creation of a micromodule on smart and sustainable mobility, with a short duration, to allow any student to learn concepts on this topic through a project-based learning overview as well as to incorporate a permanent learning process for adults.

Finally, experiences will be presented by students who develop research in the field of transport systems, which constitute opportunities and teaching-learning experiences with the involvement of research methodologies.

5.2.1 The Course 'Energy, Mobility and Transportation'

'Mechanical engineers can be at the forefront of developing new technology for energy, environment [...] transportation, safety, [...]', as advocated in the Report of the American Society for Mechanical Engineers "2028 Vision for Mechanical Engineering" [5]. This was the motivation for the creation of the course 'Energy, Mobility and Transport' under the scientific area of mechanical engineering. It is a curricular unit that aims to highlight contents and contributions from the areas of thermodynamics, thermal machines and thermal engineering to support transport systems and technologies. It is considered that its curriculum, especially as it is a curricular unit that integrates several Master courses, must have an open character to be able to integrate elements from both research on this thematic area and the articulation between personal reflection and research expertise.

This course was created in the Department of Mechanical Engineering of the University of Aveiro as an optional subject in the academic year 2007/2008, to fill in the absence of this thematic area. The curricular unit is developed in the format of theoretical and practical classes, with 2 hours per week for each type, and corresponds to six ECTS. Part of the enrolled students come from other fields (rather than Mechanical Engineering), which can range from Biology, Meteorology, Oceanography, Physics, Environmental Engineering, Civil Engineering, Electrical Engineering, Engineering and Industrial Management, Renewable Energy Systems, among others). This particularity leads to an interesting heterogeneity and distinct academic and professional experiences but poses a challenge and a higher demand for the Professor so that all students understand the subjects without losing the scientific rigor of the contents. Furthermore, this fact requires some teaching flexibility, always considering the balance in the selection of subjects as well as the development and deepening to be given to each one of them, in a scientific and technical domain in constant and rapid evolution.

With this course, it is intended that the student uses the skills acquired in previous courses (namely, in the field of thermodynamics and fluids) in understanding the transport sector, as a key sector regarding fossil energy consumption and the influence on the energy matrix of a city/region/country. The course also aims to help students acquire an enlightened attitude on how to:

- Select and correctly apply a quantitative methodology, to present, interpret and critically evaluate the results obtained;
- Recognize the importance of rigor in analysing and interpreting data in transport systems;

- Demonstrate a constructively critical approach, creativity and an attitude oriented towards incorporating research methodologies into professional activities, on an assertive and pragmatic basis;
- Identify and demonstrate strategies for achieving learning objectives.

This curricular unit has as its main objective the development of skills in the energy-environmental aspect of the transport sector. On the other hand, it has several specific objectives, in which students, at the end of the course, should:

- Be able to understand the current situation in the transport sector from an energy and environmental point of view;
- Identify and know how to use the main techniques and methodologies, both numerical and experimental, of data related to the performance and impacts of the transport system, as well as the practical applications of these tools in engineering problems;
- Develop skills in using modelling tools for pollutant emissions and fuel consumption, in order to know how to apply them in engineering studies;
- Develop knowledge about urban mobility challenges and new paradigms, with the specific objective of identifying solutions for sustainable mobility from an energy and environmental point of view;
- Have developed skills in the field of identification, characterization and application potential of intelligent transport systems and services (ITS);
- Identify the best potential for the application of alternative fuels and alternative propulsion technologies in the transport sector.

In the National Convention on Higher Education of March 15, 2019 (held at the University of Aveiro) it was mentioned that the three dimensions of the teaching of the century. XXI (the three 'Ms') are: Motivate; Mediate; and, in the end, Measure. The author of this chapter has the conviction that, in the training of students and future engineers, it is always necessary to inspire assertive observation, an autonomous posture and rigor in analysis and discussion, in addition to other characteristics. Thus, in addition to the acquisition of specific skills, it is generally intended that the student internalize and apply the following behaviors and attitudes:

- To develop skills in teamwork, time management and knowing how to communicate effectively in public;
- To develop a sense of ethical and deontological responsibility;
- To develop a proactive attitude in your learning and development;
- To acquire critical thinking and decision-making skills;

- Reflect and learn through contact with people with different experiences and perspectives;
- Develop a sense of lifelong learning.

The objectives established for this curricular unit are consistent with its syllabus, namely in the identification and understanding of the themes to be addressed in the curricular unit at the level of transport systems and in the knowledge of the necessary methodological tools to establish the link between theory and practice. There is also an interconnection between the syllabus, the defined objectives and the considered bibliography.

This course has had, over the years, an improvement with the adaptation, refinement or the introduction of new teaching methodologies and thematic updates, since the areas of energy and transport are multidisciplinary and constantly changing. This also poses a challenge for the Professor to permanently update the digital content provided to students, throughout all the years in which the curricular unit was taught. For this reason, the recommended core bibliography is primarily composed of technical reports from the European Commission, European Environment Agency (among other institutions) [2–8], and not just books dealing with this theme, since the latter does not cover all the subjects taught and may suffer from some outdated content. Still, some complementary books are suggested for each thematic chapter.

The option was to give a particular focus to the road mode since it is the main responsible for the energy-environmental imbalances in the transport sector and for the problems of traffic congestion. Nevertheless, air and rail modes are mentioned, namely through the calculation of energy consumption and emissions and emphasizing its importance and framework in the organization of transport systems.

The course syllabus has a theoretical and a practical component and includes several chapters, each of which may cover one or several classes. At the beginning of each chapter, students are provided with a list of the recommended bibliography, which includes reports, book chapters and relevant scientific papers, which must be in sufficient number to allow contact with different approaches, strategies and opinions. It should not, however, be so extensive as to inhibit the student from seeking documentation.

Each theoretical class starts with a brief recap of the material covered in the last lecture or with an interactive discussion on a question formulated at the end of the previous lecture. At the beginning of each class, the summary and the specific objectives to be achieved are displayed. The theoretical program was divided into six 'blocks' or 'pedagogical modules', which are, in turn, divided into topics.

Regarding practical classes, each one starts with the presentation of the task(s) to be developed in the class and its interconnection with the matter covered in the respective theoretical class. Once again, at the beginning of each class, the summary will be presented and the specific objectives to be achieved will be explained. In the practical component, each chapter will correspond to interdependent themes of a practical proposal for the development of a project (under a project-based learning perspective), calculation exercises or a group discussion activity for students.

The structure of the theoretical program is summarized below:

The first part of the program consists of an introductory, motivation and contextualization, necessary to frame the students on the themes to be taught to the specificities of the transport sector and its interdisciplinarity with other scientific fields (namely, energy). It is also intended that this part is particularly motivating for students' interest in the course. The student should receive information about the specificity and scope of the subjects that will be covered. A vision of the problems arising from the transport sector are introduced, namely: global energy consumption patterns and the influence of the transport sector; environmental impacts of energy (giving particular relevance to the issue of climate change). Statistics are presented at different scales, namely worldwide and in Europe.

The second part is dedicated to the presentation of the diagnosis of the transport sector, both in the global context and at the European level. It will address the evolution over time and space of the different variables that reflect the mobility of passengers and goods (such as motorization rate, modal split, among others). After the presentation of the current overview, the main challenges currently facing the transport sector are presented and described: 1) congestion and 2) energy and environment. Finally, the external components of the transport sector, which make a difficult sector to control and manage, will be addressed.

The third part includes the main aspects of the formation of pollutants in road vehicles. In this topic, the types of emissions from vehicles equipped with internal combustion engines are systematized and a comparative analysis is made between the emissions of a vehicle with a gasoline engine or with a diesel engine, separately for each pollutant. Next, the main phenomena of the production of pollutants are described. The treatment of these gases at the level of the catalyst, particulate filter and exhaust gas recirculation are also addressed, through the explanation of how these emission control systems work. Examples of engine emission maps are presented. Finally, the importance of studying the energy consumption and emissions of pollutants from road vehicles in engineering projects is exemplified, in different contexts, as well as the differences between the emission standard measurements carried out in the laboratory and real-world, on-road measurements.

The fourth part is dedicated to the presentation of the main methodologies for quantifying the impacts of transport systems. Particular attention is given to the quantification of traffic congestion, fuel consumption, pollutant emissions and noise levels. A summary approach of the main experimental monitoring systems and some numerical models used to quantify these impacts include the demonstration of several application examples. Finally, the relevance of life cycle assessment as a fundamental method of comparison between different propulsion alternatives for the transport sector is presented, using principles of energy systems analysis.

The last chapters (fifth and sixth) are dedicated to reflecting on the importance of changing transport systems, the new existing paradigms, in order to reduce energy and environmental impacts. Topics related to: 1. the importance of behavioral change; 2. the role of technology as a factor of evolution and change. In this context, the fifth chapter describes essential issues regarding sustainable mobility measures that can be implemented, the use of intelligent transport systems and services ('Intelligent Transportation Systems' – ITS) and the digitalization of the sector. Emerging themes of connected and automated vehicles, as well as mobility-as-a-service are addressed. Finally, the main technologies on the market in terms of alternative fuels and alternative propulsion vehicles will be discussed in the sixth chapter.

During the theoretical component, whenever possible, the use of practical examples will be used, namely through the exhibition of short videos (from the European Environment Agency, among other institutions) and the distribution of elements of the media about the transport sector or recent scientific articles that are relevant to the understanding of the subject in question.

The practical component corresponds to the development of practical projects, calculus exercises and student discussion activities in groups in class. It is important to convey to students the idea that the word 'practice' does not indicate a change in subject matter, but rather a change in their activity in the classroom.

The development of the project has, as the main objective, that students learn how to quantify the vehicles energy consumption and emissions in a particular case study. The topic chosen by the students can be inserted in one of three thematic areas: 1) Study of the energetic and environmental impact of a route, through the comparison of different modes of transport (road, rail, air or a combination of them); 2) Study of the fuel consumption and emissions of a fleet; 3) Study of the implementation of sustainable mobility measures in a given area. It is intended that students use the methodology 'EMEP/EEA air pollutant emission inventory guidebook – Exhaust emissions from road transport' to calculate consumption and emissions, namely through the COPERT model (a reference model for the European Environment Agency concerning road vehicles fuel consumption and emissions modelling) [9]. It is intended that, during classes dedicated to learning how to use this model, students test the methodology and case studies on their own computer (or in computer-equipped rooms), in a 'hands-on' logic, to have a greater internalization of the methodology through practical activity.

This practical work can be structured in three phases. The first phase consists of data inventory on the chosen topic, with an inventory of input parameters for the methodology, including: country; fuel characteristics (gasoline, diesel and LPG); room temperature; length of trip; characteristics of the fleet to be considered; average speed of vehicles, by type of road; percentage of circulation on each type of road. The second phase comprises the compilation, treatment and processing of data, the quantification of consumption and emissions of pollutants with computational support (namely, through COPERT) and the analysis of the results obtained. Finally, in the third phase, a final report is prepared with the work carried out. This report will be presented orally by the various working groups (maximum of three students per group) and discussed by the whole class in the last class. In a parallel activity, a demonstration of on-board emissions monitoring system will be made to the students (Figure 5.1), in order that they can understand the modelling specifications and assumptions.

Regarding group discussion activities, it is intended that students develop a topic in groups of two to three students and, finally, present it to the class. It is intended that students internalize the theoretical concepts through the presentation of case studies and their discussion, through a dynamic interaction between students and Professor. it is essentially intended that students are able to organize information on the subject in a short period of time (maximum 1 hour) and present the main findings of your research and discussion to the rest of the class. The aim of these tasks is to promote group discussion and students' communicative and expository skills. Since the communication skills of each one can be worked on and developed, it is also intended that students train their ability to communicate in public, something that is increasingly valued by companies or other professional environments (translating into participation in meetings and/or presentations to directors, employees, clients or members of a consortium). In these practical classes,

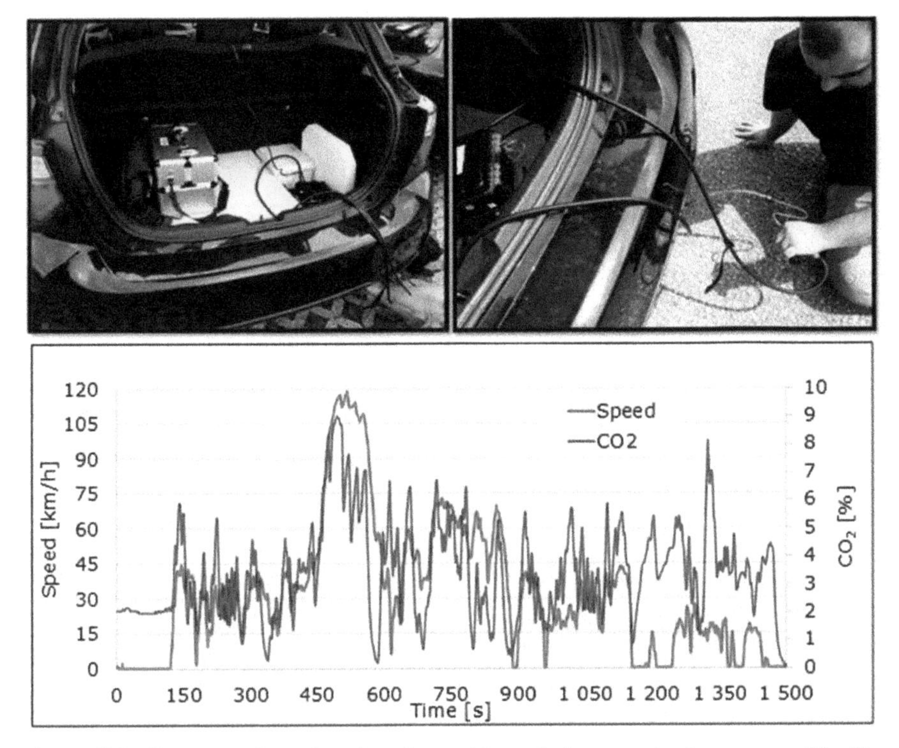

Figure 5.1 Demonstration of on-board portable emissions monitoring system (Credits: Margarida C. Coelho).

discussion forums are on European Commission regulatory documents (such as the Transport White Paper or the Smart and Sustainable Mobility strategy), on sustainable mobility measures implemented in various cities, as well as on the strategies of different manufacturers in terms of market penetration of alternative propulsion vehicles.

In preparation for one of the group discussion activities, students are encouraged by the Professor to assess a mobility challenge in their cities, which can consist of evaluating: 1) cycling mobility; 2) the supply and demand for public transport; 3) road conflicts between pedestrians and cyclists; 4) the mobility patterns of employees, students and professors at the HEI; 5) the safety of pedestrians and cyclists; 6) bicycle sharing systems; 7) of the transport of goods, including the loading and unloading process. This activity was inspired by a group work experienced by the author in a summer school organized by the network CIVITAS, where professionals from different domains (universities, Municipalities, companies) worked in groups to solve a mobility-related issue in the city of Malaga (Figure 5.2).

Figure 5.2 Inspiration gathered by the Professor from the experience taken in a summer school to an Assessment Activity for the students (Credits of the pictures: Margarida C. Coelho (left hand side); students of Energy, Mobility and Transportation (right-hand side)).

During the semester there are several lectures or seminars given by experts in the field of transport systems and technologies. Examples of these sessions are demonstrated in Figures 5.3 and 5.4. with speakers from company Toyota Salvador Caetano, North Carolina State University (NCSU) from the USA, the Technical University of Delft (TUDelft, The Netherlands) and the University of Salerno (Italy) have already participated in these sessions – this last participation took place in the scope of an ERASMUS+ mobility mission (a professor from the University of Salerno, specialist in traffic noise, taught a theoretical and a practical class of the curricular unit 'Energy, Mobility and Transport'). These sessions can take place in person or by videoconference and are scheduled for the class schedule, according to the availability of those involved.

When it is feasible to schedule visits to companies in the transport and automotive domain, the students will also benefit from this experience (Figure 5.5).

Regarding teaching methods, expository and interrogative methods will be used and, whenever appropriate, various techniques of the active method ('Active learning') should be also used, namely 'Peer learning', 'brainstorming' or 'brainstorming', case studies and project-based learning. The teaching methods to be used are preferably based on a perspective of constant interaction with students and are listed below:

- Lecture of theoretical material, followed by discussion by students on the topic(s) presented in the classroom;

a)

b)

c)

Figure 5.3 a) and b) Seminars with mobility experts (by videoconference and on-site); c) demonstration of hybrid and electric technology (Credits: a) Gonçalo Correia; b) and c): Margarida C. Coelho).

Figure 5.4 ERASMUS+ mission partnership between the Professors from Italy and Portugal: a) lecturing noise concepts and measurements methods; b) noise levels experimental monitoring in the university campus using Noise Capture app (Credits of the pictures: Left-hand side: Margarida C. Coelho; Right-hand side: Noise capture app by IFFSTAR).

Figure 5.5 Visit to an automotive company (Credits of the pictures: Margarida C. Coelho).

- Practical calculus exercises;
- Hands-on classes to use numerical models to quantify the energy and environmental impact of a given fleet or route;
- Forums for discussion of current and relevant topics for the transport sector;
- Seminars by invited experts from the energy and transport sector.

The methodology of the course includes a conceptual approach and a broad contextualization to current issues, to meet the objectives of the course. Critical issues, priorities and challenges in the transport sector are analyzed and quantified. The Professor should bring his/her expertise to the classroom regarding the 'hot topics' of research in the field, which presupposes a continuous evolution of thematic contents.

On the other hand, achieving the learning objectives involves analyzing and discussing practical cases. Thus, students are encouraged to apply the acquired knowledge to application examples, with particular emphasis on the management and impacts of transport systems.

Thus, the habit of participating in classes is encouraged from the beginning of the course. The first classes, as they address general concepts, some of which are already known by the students, may constitute privileged spaces for dialogue between the Professor and the students, as well as among the students themselves. These 'discussion forums' should be conducted by the Professor, to inspire the habit of participating in students. Finally, to close most of the programmatic issues, several concrete case studies that illustrate the subject at hand will be analysed with the students in an interactive way. These should be presented to stimulate students' curiosity and encourage their ability to put forward hypotheses for explanation and to equate potential solutions, which should be analysed rigorously and critically. Students should find, for themselves, the answers to the questions formulated by themselves or by the Professor, who can help the further discussion.

The choice of practical examples with integrated approaches, which include areas of intervention of future professionals and situations that require the participation of specialists from various areas (with which future professionals will have to interact and work), is particularly relevant and determinant in the motivation of students and should carefully carried out by the Professor. On the other hand, and whenever deemed convenient, the Professor should approve the exposure by students of knowledge already acquired in the context of previous classes of the course to the detriment of her own exposure. However, at the end of the exhibition, the Professor should clarify and summarize the concepts expressed.

The timely distribution of reliable bibliography on the discussed subjects is crucial. Additionally, it is considered particularly important to confront students with different and even contradictory, opinions regarding the same topic. In this way, their critical spirit can be developed, through the recognition of the existence of different interpretations and points of view regarding scientific matters that are sometimes presented or explained as constituting an absolute truth. However, it should be Professor's duty to adequately clarify and frame the different points of view and in the end, carry out a clarifying summary of the situation, so that students do not feel lost between contradictory positions. The presentation of the material by the Professor and the starting point for the interactive spaces within the class will be accompanied by the most appropriate means for each case, including the projection of short videos (particularly from the European Commission and from the European Environment Agency), use of the whiteboard and the multimedia projector, as well as sharing news from the written press and other elements relating to the issue of energy, mobility and transport.

In the practical component of the course, students will be faced with situations that they might have to face in their professional practice in the mobility field. The introduction of a project within the scope of the course results from the perception that the preparation of a written report that synthesizes the relevant information on a given subject and the training of an oral presentation is very useful for the training. This component is of particular importance to developing students' ability to search for and select appropriate information, as well as their self-learning skills. It is intended that the student learns to approach and develop the chosen theme in a logical, structured and coherent way and that he/she also learns to interpret data related to the quantification of energy consumption and emissions from the transport sector. The learning of techniques that you can use in your future professional will be privileged during the practical component. Classes with a practical component will accompany, whenever possible, the themes covered in the theoretical component, so they will be properly spread with classes reserved for bibliographical research and discussion of the topic under development by students within the scope of their practical project.

As mentioned, the main objective of the project component is to stimulate the students' self-learning capacity. Therefore, the Professor should intervene as little as possible in the work to be carried out during the semester but should provide advice that can be useful to students in their search for information and should clarify their doubts whenever they request their intervention. However, the clarifications provided must be made in a way to help find the explanations and not to directly provide the results. At the end of the semester, there should be a public presentation of the projects, to which all students must attend. At the end of the presentation, they will carry out a self-assessment of their work and try, as far as possible, to assess whether the students have internalized the phasing of an energy-environmental analysis project for the transport sector.

At the beginning of the project, supporting documentation is distributed (through the e-learning system), which includes: a brief introduction to the topic; the goals to be achieved; an example guide to the structure of the report; a description of the methodology and techniques to be used, sufficiently detailed to allow the work to be carried out in an almost autonomous way; a list of the recommended bibliography. The Professor should briefly and objectively explain the work to be carried out, highlighting the specific objectives to be achieved. In order to facilitate the execution of the work, working groups of two to three elements (maximum) should be set up.

The project must follow the structure of a technical-scientific report. In addition to training in reporting, this practice will contribute to the systematization and consolidation of acquired knowledge. Students should carry out bibliographic research on the subject in question, in addition to the documentation provided, using the means at their disposal in databases, the internet, books, scientific and technical journals. The overall information must be summarized in a final report. At the end of the semester, students must make an oral presentation of the subject, lasting 10 minutes. The oral presentation will be followed by a period of 5 minutes of discussion, during which the Professor will ask questions about the topic discussed, to clarify any doubts raised when reading the report and during its presentation.

5.2.2 Micromodules

The European Consortium of Innovative Universities (ECIU University) is a network of 13 universities united since 1997 that desires to create a new educational model on a European scale. The ECIU University gathers learners, Professors and researchers to cooperate with cities and businesses and solve real-life challenges. The ECIU University consortium wants to be the first European group of universities where learners, Professors and researchers cooperate with cities and businesses to solve real-life challenges. Also, it wants to create, test and evaluate a whole new educational pedagogy. This will lead to focus all the university activities (from education, research, administration and support) through innovation and valorisation.

Micromodules are a key tool within the concept of teaching and learning that the ECIU University wants to promote. Its definition (assumed by ECIU University) is 'A micromodule is a short learning experience that is formally assessed and supports learners to fill their knowledge gaps and boost their capabilities in order to successfully engage in ECIU University challenge-based activities'.

The first 3-year phase of the ECIU University focuses on the UN Sustainable Development Goal 11 (ODS11) 'Sustainable cities and communities'. Thus, micromodules were identified that could be taught online, in English, that accomplish the criteria of 1– 3 ECTS) and, finally, could be framed in at least one of the following main topics: 'Energy and sustainability'; 'Sustainable urban environments'; 'Resilient communities'; 'Transversal Competences'; and 'Transport and mobility'.

The 'Smart and sustainable mobility micromodule' will allow students to understand the dynamics of the transport sector, namely in terms of the evolution of the automotive fleet and mobility patterns, use of different means of transport, evolution in the use of fuel, among others. Some metrics and some mobility performance indicators will be worked on so that students can assess the efficiency and sustainability of the sector. The topics related to measures, policies, challenges and existing paradigms (including at the technological level) will be crucial for the development of skills to identify opportunities for smart and sustainable mobility.

The methodology of the micromodule includes a conceptual and contextual approach extended to current topics to meet the objectives of the course. The achievement of learning objectives involves the analysis and discussion of specific case studies. The critical issues, priorities and challenges of mobility are analysed, discussed and quantified.

5.2.3 Incorporate Students in Research Activities

The integration of bachelor or master students in research activities is a strategy for their motivation and involvement. The Professor should promote, whenever possible, the possibility for students to participate in meetings of the research group, to join the support team for the organization of courses/conferences, thus how to participate in ongoing project activities, such as: experimental monitoring campaigns (which consist of the quantification of traffic and energy-environmental performance variables in the transport sector) and numerical modelling or simulation related activities, as shown on Figures 5.6 and 5.7.

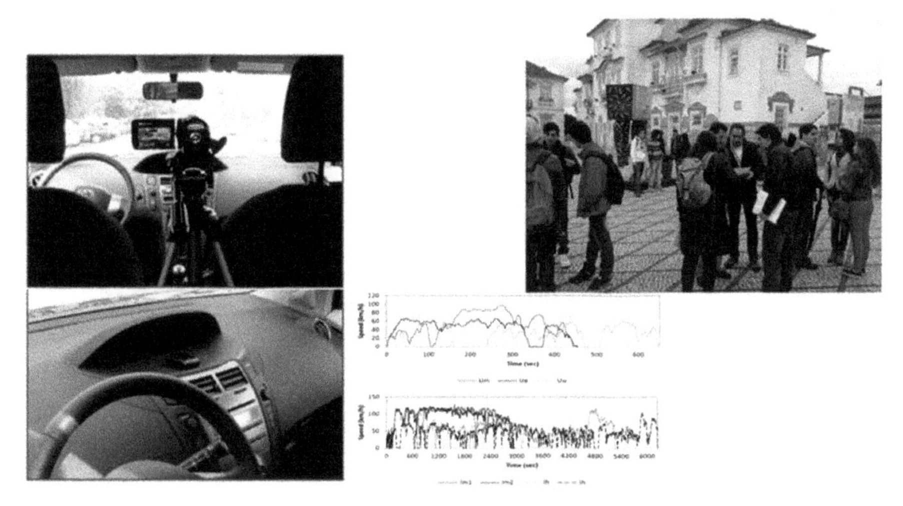

Figure 5.6 Participation of bachelor and Master students in research activities (such as experimental monitoring campaigns) (Credit of the pictures: Margarida C. Coelho).

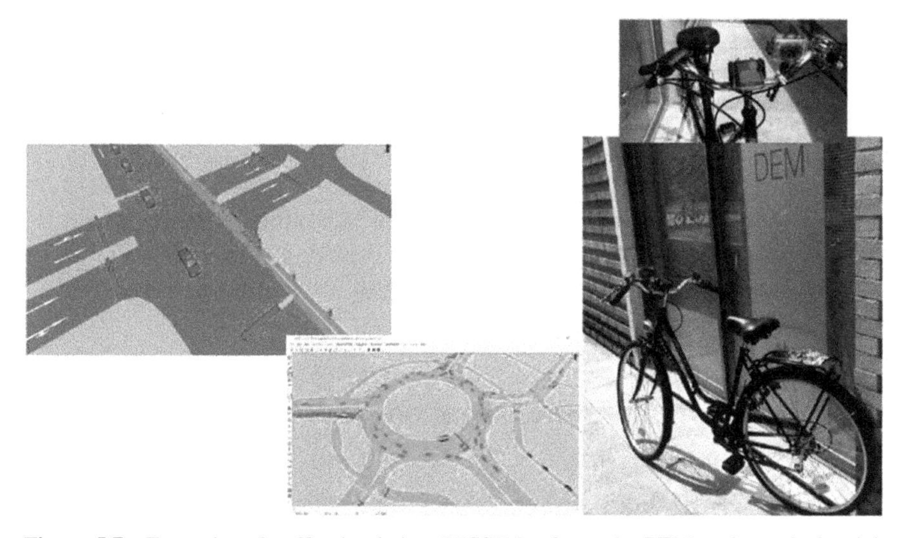

Figure 5.7 Examples of traffic simulation (VISSIM software by PTV) and practical activities with an instrumented bicycle applied on Master dissertations (Credits: Luis Campos and Joana Guimarães, students of Mechanical Engineering Master degree).

If the work developed by the student has the possibility to be submitted to a conference of the transport field, this is emphasized by the Professor. It is very rewarding for the student to present his/her research in a scientific

Figure 5.8 Participation of Master students at international conferences on mobility and transportation, where they have the possibility to interact with senior researchers: example of the prestigious Transportation Research Board Annual Meeting, (Washington DC) Credits of the pictures: Margarida C. Coelho.

venue and get feedback from senior and well-known researchers (Figure 5.8). Some of these works are presented at important conferences in the field of transport or even published in international scientific journals, which contributes not only to the dissemination of their work in scientific forums, but also for students to participate in the writing of a scientific article and/or submit a communication in English.

5.3 The Power of Internationalization Under the Teaching Process

The themes of energy and transport have a very strong global dimension. Thus, the topics contribute to the integration of international and intercultural dimensions in the students' curriculum, even by pedagogical strategies of 'Internationalization at Home'. It is a privileged opportunity to implement the relationship between research and teaching-learning in response to the societal challenge 'intelligent, ecological and integrated transport', internationalization in the classroom (through the presence of experts and the sharing of results of international projects), the use of information and communication technologies, as well as the promotion of closer ties between the companies and HEI. Since the subjects are interdisciplinary, there is a natural predisposition for the integration of results and international research practices in the students' curriculum. Thus, the ultimate objective of this approach is to encourage critical reflection and a scientific spirit in the different topics covered, through the articulation between the transport-related topics and the RandD activities of the Professor. It is intended with these activities to implement a coherent posture in the set of teaching-research activities.

In addition to Portuguese students, other students of different nationalities attended the 'Energy, Mobility and Transport' course, namely students from Germany, Brazil, Slovakia, France, Netherlands, Indonesia, Iran, Italy, Mozambique, Norway, Poland, São Tomé and Príncipe, Sweden and Venezuela. The academic year 2019–2020 was an example in which eight different nationalities were present in the class; this fact, added by the existence of students with different basic education, makes teaching challenging and stimulating. When this happens and students do not come from countries where Portuguese is the official language, the Professor promotes face-to-face teaching in English in the theoretical class and there is a practical class in which the practiced language is preferably English. Additionally, all study materials are written in English.

The possibility of sharing international and multicultural experiences on the themes of Energy and Transport is always enriching for the Professor. It becomes possible to interpret the information produced in different cultural contexts and recognize the impact of these same differences on the teaching-learning process. This particularity leads to an interesting heterogeneity and distinct academic experiences, but it poses a challenge and a higher demand for the Professor so that all students understand the subjects without losing the scientific rigor of the contents. Furthermore, this fact requires some flexibility in the form of teaching and requires a balance in the selection of content and the selection of development to be given to each one of them. Thus, the final objective of this approach is to encourage critical reflection and discussion of the different topics covered in the course, as well as the students' own experiences in their country of origin.

On the other hand, it is also intended to promote the development of skills that allow students to work in multicultural environments, as the labour market is increasingly globalized. Thus, whenever possible, students are encouraged to work in mixed nationality groups; when the work presupposes an oral presentation, the use of English is encouraged.

Second, several methodological tools were used to include international perspectives on the topic of mobility. For example, the application of technology has been done through video conferences with professors from European and US higher education institutions. Online research is also encouraged on certain content in different international contexts and results of RandD projects (which often involve partners from European and US institutions) in which the professor has participated are presented.

Third, papers published in relevant international journals or conferences in the field of transport and mobility are presented in class, which helps students to participate in the reading and understanding of a scientific article written in English. The themes of the course itself are interdisciplinary and contribute to the integration of results and international research practices in the students' curriculum. Thus, the final objective of this approach is to encourage critical reflection and a scientific spirit in the different topics covered, through the articulation between the thematic contents of the curricular unit and the effective scientific competences of the Professor. Above all, it is intended that all students understand the subjects without losing the scientific rigor of the contents. Thus, we seek to include the application of the scientific method in the theme of mobility, namely through the presentation of objectives, methodologies, methods and results of national and international RandD projects in which the Professor participates.

Finally, it was already mentioned that a cycle of seminars is organized in each academic year, in which professors or researchers from several European and US HEI in the field of transport systems participate. They share the results of ongoing projects as well as disruptive themes in the transport sector, in person or by videoconference. The curricular unit is also a stage for the exchange of good practices and experiences through ERASMUS+ mobility missions.

5.4 Conclusion

In this way, and taking into account the HEIs commitment to serving science and society, the pedagogical adequacy of the curricular unit to the interests and needs of the institution's project, the scientific and business communities of the region and at the level of the national strategy. The inclusion of Mobility and Transport related topics in curricular units, micro modules and Master dissertation topics represents an advance in the perception of the relevance of the transport sector from an energy perspective, namely because it is a very difficult sector to control and constantly changing. Also, it is of utmost importance to implement a coherent posture in the set of teaching-research activities, to develop research-oriented learning activities, because students involved in research activities will be motivated in the learning process.

5.5 Acknowlegdements

This work was supported by the projects UIDB/00481/2020 and UIDP/00481/2020 - FCT - Fundação para a Ciência e a Tecnologia; and CENTRO-01-0145-FEDER-022083 - Centro Portugal Regional Operational Programme (Centro2020), under the PORTUGAL 2020 Partnership Agreement, through the European Regional Development Fund; DICA-VE (POCI-01-0145-FEDER-029463) project funded by FEDER through COMPETE2020, and by national funds (OE), through FCT/MCTES. The author acknowledges her reserch team members, C. Guarnaccia (University of Salerno, Italy), G. Correia (TUDelft, The Netherlands), Toyota Caetano Auto and Renault CACIA for the cooperation in seminars and visits given to the students of "Energy, Mobility and Transportation".

References

[1] J.-P. Rodrigue, 'The Geography of Transport Systems', New York: *Routledge*, 456 pages. ISBN 978-0-367-36463-2, 5th Edition, 2020.

[2] EC, 'Sustainable and Smart Mobility strategy', Communication From The Commission To The European Parliament, The Council, The European Economic And Social Committee And The Committee Of The Regions Sustainable and Smart Mobility Strategy – putting European transport on track for the future, European Commission, COM/2020/789 final, 2020.

[3] EC, 'Energy and Transport in Figures', European Commission, 2021.

[4] EEA, 'TERM: Transport and environment reporting mechanism', *European Environment Agency*, 2020.

[5] ASME, '2028 Vision for Mechanical Engineering: A report of the Global Summit on the Future of Mechanical Engineering', 2008, ASME, New York, USA.

[6] ICCT, 'White paper pathways to decarbonization: the european passenger car market in the years 2021–2035', *International Council on Clean Transportation*, 2021.

[7] TOI, 'Changes and Challenges in Future Transport – Drivers and Trends'. *Institute of Transport Economics*, 2021.
[8] UNECE, 'Recommendations for Green and Healthy Sustainable Transport – "Building Forward Better", United Nations, 2021.
[9] Emisia, 'COPERT: Computer programme to calculate emissions from road transport', 2021.

6

Push and Pull: Sustainability Education for 21st Century Engineers

Salma Shaik, Lakshika N. Kuruppuarachchi, and Matthew J. Franchetti*

The University of Toledo, 2801 W. Bancroft St, Toledo, OH – 43606, USA
E-mail: salma.shaik@utoledo.edu; lakshika.kuruppuarachchi@utoledo.edu; matthew.franchetti@utoledo.edu
*Corresponding Author

Abstract

Though technological advancements are reaching dizzying heights, the world is still struggling with dependence on fossil fuels, societal and economic issues that have made our societies and planet unsustainable. It is critical that sustainable development (SD) principles are adopted in full swing by one and all to minimize the use of non-renewable resources, reduce the overall environmental impact, improve socio-economic conditions and provide access to education, healthcare and justice for all. Realizing this need, United Nations General Assembly put forth 17 Sustainable Development Goals (SDGs) to be achieved by 2030 for a more sustainable future. There has been a steady rise in interest from today's youth and industry alike to adopt sustainability principles. Education for Sustainable Development (ESD) can empower today's youth, especially engineers whose work impacts all walks of life, with the skills and values required to play a central role in steering the planet towards a sustainable and regenerative path. This chapter provides a brief history of SD and ESD, explores the push for SD from the students and pull from industry for SD adoption along with discussing required skills and ESD framework for engineers in the 21st century context.

Keywords: sustainable development, sustainability, environmental engineering, 21st century, engineers, entrepreneurial mindset, higher education, push and pull.

6.1 Introduction

The advancements in science and technology and huge industrialization led to the improvement of human societies and have been a boon to economic growth over the last century. But they have undeniably strained the earth's natural resources resulting in catastrophic environmental and climatic conditions making the possibility of a sustainable planet still a distant dream. The widening income disparities due to the unequal distribution and access of resources are creating an unhealthy imbalance that might hinder further prosperity. This problem would be further accentuated by the growing demands for food, water, shelter, transportation, economic growth opportunities, manufacturing and service industries by the increasing world population and improvements in living standards.

6.1.1 History of Sustainable Development

Since the beginning of the 21st century, there has been rising awareness about the urgent need for SD along with a growing realization and a sense of responsibility to preserve the earth for future generations. In 1987, the United Nations World Commission on Environment and Development (UNWCED) defined SD as the 'development that meets the needs of the present without compromising the ability of future generations to meet their own needs' [1]. The movement towards SD has been spearheaded by the United Nations (UN) which identified 17 SDGs and 169 targets as part of their 2030 agenda calling for immediate action to address critical issues facing humanity and the planet [2]. Research aligning with the SDGs has gained momentum over the past few years along with a growing interest in sustainability-based courses and degrees in academic institutions [3–5]. A brief overview of the key events in the history of SD is presented in Table 6.1 on page 143 [6–9].

Table 6.1 Key Events in the History of Sustainable Development

1980–1989	1990–1999	2000–2009	2010–2015
Global 2000 Report (1980): Biodiversity is recognized as critical to the proper functioning of the earth's ecosystem	*International Institute for Sustainable Development (IISD, 1990):* groundwork for SD agenda.	*UN Millennium Development Goals (MDGs, 2000):* combat poverty, disease, illiteracy, environmental and inequality issues by 2015.	*MDGs (2010):* 8 MDGs to tackle poverty rates, spread of HIV/AIDS and to provide universal primary education by 2015.
World Health Assembly (1981): Global strategy adopted to ensure quality of life and health.	*UN Summit for Children (1990):* harmful effects of environmental degradation on future generations are recognized.	*World Summit on SD (2002):* promoted partnerships as a non-negotiated approach to sustainability.	*Rio 20 (2012):* agreements on smart measures for clean energy, jobs and more sustainable use of resources.
The UN World Charter for Nature (1982): Raised awareness about unhealthy dependence on natural resources	*Global Environment Facility (1991):* decision-making power to developing countries to address local sustainability issues	*Global Reporting Initiative (2002):* guidelines for reporting on sustainability dimensions of business activities are established.	*World Federation of Engineering Organizations (WFEO, 2013):* Model Code of Practice and Interpretive Guide for engineers to implement SD.
Third World Network (1984): to raise awareness on sustainability issues	*Earth Summit (1992):* agreements on preserving biological diversity, climate change and desertification	*Kyoto Protocol (2005):* guidelines to legally bind countries to GHG emissions reduction goals.	*SDGs for 2030 (2015):* to work towards a more sustainable future.
Climate Change (1985): buildup of 'greenhouse gases' in the atmosphere are discussed.	*World Trade Organization (WTO, 1995):* established with a formal recognition of the linkages between trade, environment and progress.	*Millennium Ecosystem Assessment (2005):* scientific information on the effects of ecosystem changes on humans.	*Paris Agreement (2015):* to limit global warming to below 2 and to achieve a climate neutral world by mid-century

Continued

Table 6.1 Continued

1980–1989	1990–1999	2000–2009	2010–2015
Brundtland Report (1987): weaved together sustainability aspects and global solutions popularizing the term 'SD'	*World Summit for Social Development (1995):* a consensus reached to eradicate absolute poverty	*Svalbard Global Seed Vault (2006):* to preserve the genetic diversity of the world's food crops for future generations.	*Wild for Life (2016):* campaign to protect endangered wildlife species.
Montreal Protocol (1987): to limit the use of substances harmful to the Ozone Layer.	*4th World Conference on Women (1995):* recognized that there still are obstacles to the realization of women's rights as human rights.	*Green Economy (2008):* rising investments in environmental actions, and a low-carbon economy leading to the growth of 'Green Economy'.	*Minamata Convention on Mercury (2017):* provisions to protect human health and environment from emissions and releases of mercury compounds.
Intergovernmental Panel on Climate Change (IPCC. 1988): to assess and co-ordinate climate change research.	*World Water Forum (1997):* raised political and social awareness about the importance of water.	*G20 Pittsburgh Summit (2009):* measures to phase out fossil fuel subsidies, sustainable consumption, and targeted support for poorest people	*Decade of Ecosystem Restoration 2021–2030 (2019):* aims to scale up the restoration of degraded and destroyed ecosystems to fight climate crisis

6.1.2 Education for Sustainable Development (ESD)

ESD takes the center stage in the race to achieving SDGs by 2030 and The United Nations Educational, Scientific, Cultural Organization (UNESCO) has called upon academic institutions to make ESD a core component in all educational offerings by 2025 [10–13]. ESD is defined as education that encourages and helps learners of all ages to obtain the required knowledge, skills, values and attitudes to address the interconnected global challenges ranging from climate change, environmental degradation, loss of biodiversity to poverty and inequality [12, 14]. Current and future generations should be well prepared to be responsible custodians of our planet and hence educating our youth is the best option to protect the earth and to sustain human societies. Identifying this need, over 1,000 university presidents and vice-chancellors had signed the Halifax Declaration, Swansea Declaration, Copernicus Charter, Talloires Declaration, Kyoto Declaration and Lunenburg Declaration at the end of the 20th century committing their institutions towards a sustainable path in all their endeavors [15, 16]. Beyond 2020, UNESCO is continuing to strive towards further scaling up ESD across all educational offerings of the world [17]. Keeping in pace with the progress of SD initiatives over the years, ESD has evolved to play a central role in the implementation of SD and is recognized as a key contributor to the 2030 Agenda as outlined in Figure 6.1 on page 145.

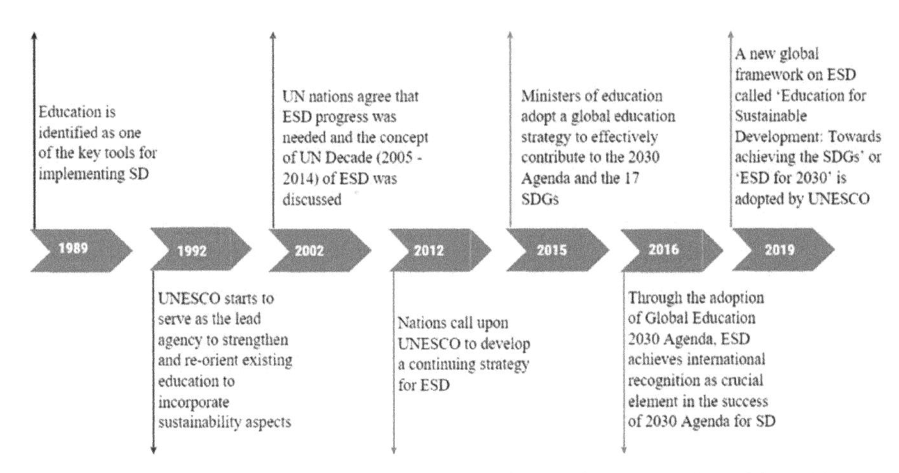

Figure 6.1 History of Education for Sustainable Development [Adapted from 17].

The importance of embracing sustainability education as a vital field and preparing students for a post-2030 agenda for SD cannot be stressed enough [18]. The relationship between research and action is crucial for achieving SDGs and hence ESD should generate, integrate and apply research-based knowledge to provide solutions to environmental-social problems [19, 20]. Complex social issues such as environmental quality, natural resource allocations, human equality and rights and their economic aspects can be incorporated into environmental education to impart ESD [21].

6.2 Push from Students

6.2.1 Youth Sustainability Activism

One of the heartening developments of the 21st century is the growing willingness of Millennial and Gen Z populations to embrace eco-friendly products and sustainable living more than their predecessors [22–24]. Around three-quarters of millennials are willing to change their buying habits by spending more on eco-friendly and sustainable products [25–28]. There is more awareness among today's youth about the detrimental effects of unsustainable practices which have given rise to youth-led sustainability and climate change action initiatives [11, 29]. Youth activists such as Greta Thunberg who inspired children across 150 countries to take part in the global strike, Mari Copeny known famously as 'Little Miss Flint' who brought national attention to the water crisis in Flint, Michigan and Malala Yousafzai who won the Nobel Peace Prize among many others exemplify the powerful role today's youth play in the global world [30–32]. These undertakings have garnered the attention of global media, political and corporate leaders who can no longer turn a blind eye to the severe social, ecological and environmental challenges that future generations need to face. Students are turning out to be the leading drivers of change within many establishments and access to information, increasing knowledge of global issues and degradation of the environment around them has further fueled their passion to play central roles in preserving the planet and creating a sustainable future [33].

6.2.2 Student Sustainability Initiatives

Millennials and Gen Z will continue to shape the climate action and sustainability initiatives of many governments and industries by convincing them to alter existing practices, to adopt new approaches and by creating demand for sustainable products [34]. An example in the case is from the University of Sussex where student activists were claimants on a major climate change lawsuit against the UK government about their failing 2050 carbon target which resulted in a revision of the target by the government [35]. Similarly, Melbourne University's Fossil Free campaign which was organized by students addressed the university's reservations on divestment [36]. In the US, the organizers of United Students Against Sweatshops have been successful in working towards workers' rights around the world [37].

Students from Duke University played a central role in transforming a $6 billion endowment into socially responsible investments [38]. On the other hand, Boston University students won their campaign for gender-neutral housing on campus [39].

Several other student-initiated, student-designed, and student-facilitated programs call for greater action from universities to be more environmentally friendly and partner with campus stakeholders to research, rethink, investigate and tackle the universities' greatest sustainability issues [40–46]. High school students are taking initiatives as well to combat climate change and leading their schools to reduce environmental impact [47–49]. According to an extensive survey of students in higher education around the world, 81% say that they would like to learn about SD, 91% agreed that all universities and colleges should actively incorporate and promote ESD and 93% echoed that governments should address climate change with utmost priority [50]. Some studies further reveal that students are willing to take jobs with relatively less pay from companies that are more sustainable [50, 51].

6.3 Pull from Industry

We are seeing a 'push' from students interested in making a positive difference in the world and a 'pull' from industry needing sustainability-related skillsets. Global economic and workforce trends demand engineers to be innovative, quick thinking and adaptive to fast-changing professional environments. Nowadays, industries are concerned more about their carbon footprint and attempt to preserve good sustainable practices and to increase growth and global competitiveness. According to the Bureau of Labor Statistics, in May 2020, United States employs approximately 50,260 environmental engineers. These professionals earn a median salary of about $92,120 per year. The projected job outlook for employment of environmental engineers is 3% growth from 2019–2029, change as the average growth rate for all occupations is 4% [52]. The government authorities such as the United States Environmental Protection Agency (U.S EPA) are responsible for the protection of human health and the environment. EPA implements environmental laws, monitors the environment, and provides technical support to minimize the threats and support recovery. Environmental engineers play a huge role in both public and private sectors to accomplish environmental sustainability.

6.3.1 Involvement of Environmental Engineering in Industries

Environmental engineers use the principles of science and engineering to overcome environmental issues and improve environmental sustainability, to provide a healthy environment for all living organisms. Knowledge of chemistry, biology, ecology, applicable laws and regulations is crucial when addressing global issues, such as unsafe drinking water, climate change, marine pollution, air pollution and other similar issues.

6.3.2 Environmental Engineer vs Sustainability Engineer

Sustainable engineering is a broader subject than environmental engineering. Sustainability is traditionally divided into three main parts: environmental, social and economical. A sustainability engineer is typically focused on the entire design and production lifecycle of a product, for example, developing and testing more efficient car fuels or appliances. An environmental engineer, on the other hand, is more focused on tasks such as site remediation, depletion of natural resources, the effects of pollution, and how to apply technical knowledge to address these issues, such as creating more efficient ways to recycle waste, reuse materials in new buildings or remanufacture products using existing parts.

6.3.3 Role of an Environmental Engineer

Industries expect a wide range of responsibilities from an environmental engineer. From preparing, reviewing and updating environmental investigation reports to advising corporations and government agencies about procedures for cleaning up contaminated sites. The job description may differ based on the background of the company, but some of the common duties are analyzing scientific data and performing quality control checks, monitoring the progress of the environmental improvement programs, ensuring the company compliance with environmental regulations and designing projects that lead to environmental protection.

6.3.4 Distribution of Environmental Engineers in Different Industries

Industries with the highest levels of employment in environmental engineering occupation as of May 2020 are shown below in Table 6.2 on page 149.

Table 6.2 Distribution of employments in different industries [52]

Industry	Employment	Industry Employment %
Architectural, Engineering, and Related Services	15, 020	1.00
Management, Scientific, and Technical Consulting Services	9, 960	0.65
State Government, excluding schools and hospitals (OES Designation)	6, 520	0.30
Local Government, excluding schools and hospitals (OES Designation)	3820	0.07
Federal Executive Branch (OES Designation)	2850	0.14
Waste Treatment and Disposal	780	0.79
Metal Ore Mining	220	0.53

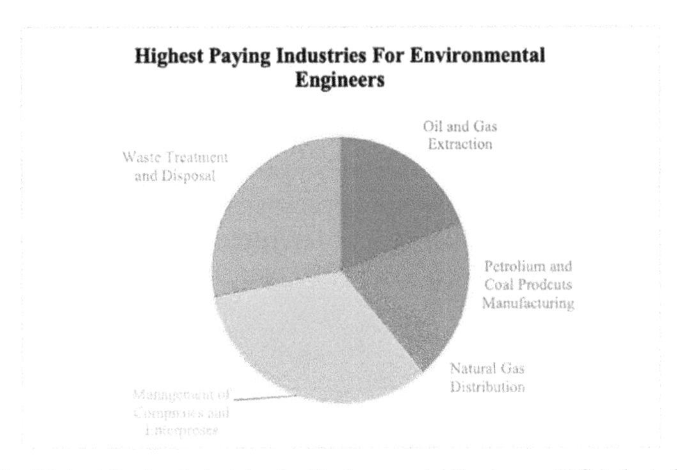

Figure 6.2 Highest Paying Industries for Environmental Engineers, U.S [adapted from 52].

Individuals passionate about conserving the environment can make their career paths in different directions, such as sustainability specialists, environmental engineers, sustainability managers, sustainability consultants, sustainability analysts and so on. All these positions have the goal of overcoming environmental issues and protecting the resources for the future.

6.3.5 Industry Sustainability Practices – Case Studies

The airline industry is heavily dependent on fossil fuel, and it accounts for roughly 2% of global carbon dioxide emissions [53]. Recently, Delta Airlines announced their commitment to mitigate all emissions – in the air and on the ground – starting March 2020 [54]. Their top initiatives include carbon reduction from their own activities and from suppliers, industry colleagues, customers, global partners, investors and stakeholders. The airline was recognized as one of America's most sustainable companies by Barron's in 2020 for its sustainability efforts such as voluntarily capping greenhouse gas emissions, replacing single-use plastics on board, partnering with bio-fuel innovators to advance the development and production of sustainable aviation fuel, etc.

New research based on a survey of 150 leading fashion and textile industries, shows that sustainability is a critical component of their operations. One of their top two strategic objectives was to implement sustainability measures for their business. The report 'Is Sustainability in Fashion?' points out that the industries are introducing sustainability measures through their supply chain from sourcing sustainably produced raw material, introducing a circular economy approach, reducing greenhouse gases, and investing in new technologies like 3D printing and blockchain [55]. Nike, an American multinational corporation, set its 2025 targets on carbon emissions reductions, waste generation, water consumption and green chemistry to protect the planet. They aim to reduce greenhouse gas emissions, through 100% renewable energy, and through increasing environmentally preferred materials, waste reduction through improved packaging, waste diversion from landfill from product recycling, 25% reduction of freshwater usage and adopting clean chemistry alternatives across their supply chain [56].

6.3.6 Environmental Engineering: A New Major

Environmental engineering is relatively a new and fast-growing major in most colleges and universities. Traditionally environmental engineering was a sub-set of Civil, Chemical and Mechanical engineering degree programs to incorporate sustainability principles in design, development, and construction initiatives [57]. Environmental engineers in a civil engineering program tend to focus on matters in air quality, wastewater treatment and solid waste and hazardous waste management disciplines. With the extensive concerns and awareness of these broad environmental issues, many universities offer a separate environmental engineering degree program or even as a separate department of Environmental Engineering.

There are approximately 98 accredited universities in North America offering degrees in Environmental Engineering [58]. 81 of these are designed as environmental engineering degrees, while others are integrated

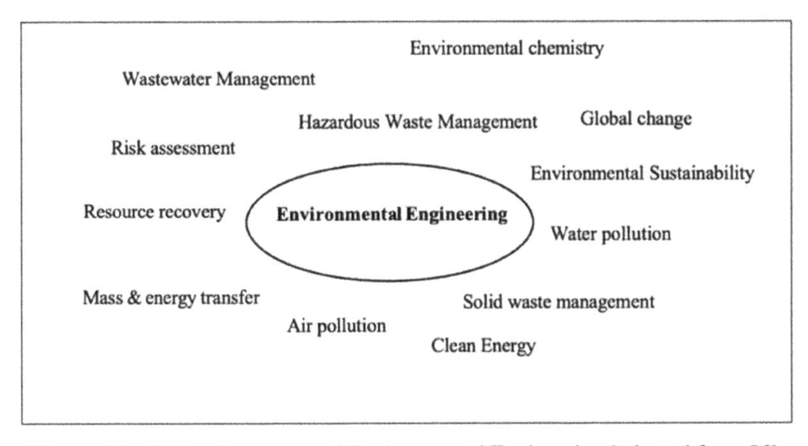

Figure 6.3 Major focus areas of Environmental Engineering [adapted from 56].

with additional disciplines such as environmental resource engineering, bio-environmental engineering, earth and environmental engineering, civil and environmental engineering technology, environmental sanitary engineering, earth system science and environmental engineering, environmental and ecological engineering, environmental engineering and science, environmental systems engineering and sanitary engineering.

6.3.7 Entrepreneurial Thinking Through 3C's

In a fast-paced world, the solutions found today may be out of date by tomorrow. Engineers with an entrepreneurial thinking mindset could change the world. Students acquire a range of engineering skills when pursuing a degree but developing a state of mind that seeks new opportunities and wants to make a difference through their work. Engineers with this state of mental capability will acknowledge the new opportunities is more essential. A curious student will integrate information from various sources to gain insight. The entrepreneurial-minded student consistently integrates the knowledge he/she gains with their own discoveries and develops innovative solutions. They can identify unexpected opportunities to create extraordinary value and continue to learn from failure.

In addition to the technical knowledge and skillset, an environmental engineer must possess an entrepreneurial mindset in today's world [59]. London et al. [60] assess the impact of using online discussions to develop these skills in environmental engineering undergraduates. They examined

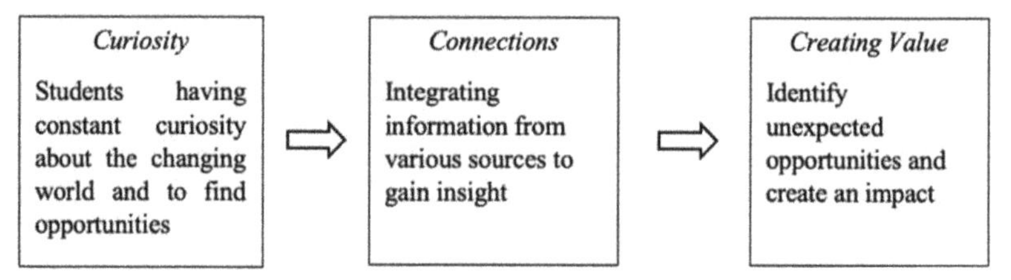

Figure 6.4 The 3C's Summarized add [adapted from 65].

how discussion prompts outside the classroom can be another effective tool. The studies [61–63] provide examples of infusing entrepreneurial mindset learning and Kern Entrepreneurial Engineering Network (KEEN)-inspired models into a set of engineering courses. Sababha et al., [64] propose a workshop integrated with the work placement graduation requirement to address the lack of an entrepreneurial thinking. The results show that the training workshop was very successful in achieving its purpose of developing enterpreneurial attitude and mindset. Adding the important dimension of entrepreneurial mindset to the training of the next generation of engineers helps to find success by creating extraordinary values for others. The KEEN partners with more than 45 universities across the United States to benefit engineering graduates with an entrepreneurial mindset, so that they can emphasize personal, economic and social values throughout their professional careers [65].

6.4 21st Century Engineer

6.4.1 Challenges of 21st Century

Myriad challenges such as environmental degradation, climate change, unsustainable food systems, solid waste disposal, depletion of renewable sources of energy, loss of biodiversity, poverty and inequality as outlined in Figure 6.5 on page 153 have plagued humanity since the dawn of the 21st century [12, 66]. To overcome these unique challenges, there is a need for drastic changes in the functioning of businesses, governments and individuals alike. These challenges continue to intensify as the global population expands toward 10 billion people by 2050 and as demands for clean water, food and energy rise, all in the context of global climate change [67]. UN identifies that the success of SD in the increasingly globalized and interconnected world would be striving towards inclusivity and progress in economic, social,

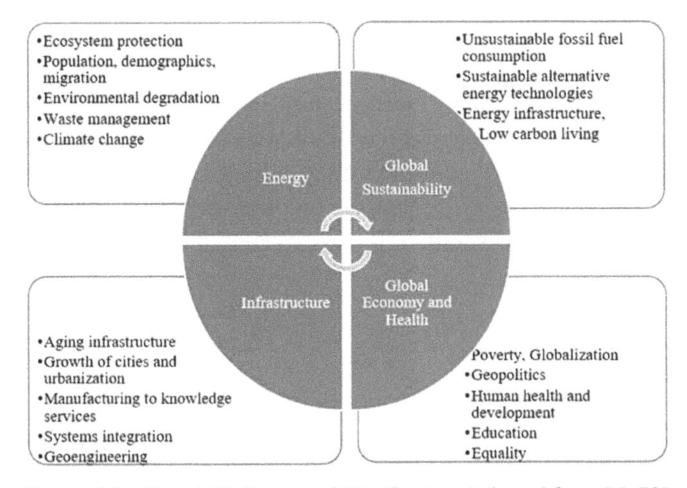

Figure 6.5 Grand Challenges of 21st Century [adapted from 75, 72].

environmental aspects of sustainability along with effective governance and peace and security [68]. Many researchers have emphasized that ESD can be a powerful tool for teaching and learning to combat and help humanity to change gears from an unsustainable way of life to a regenerative one [69–72]. ESD could equip learners with the necessary knowledge, skills, values and attitudes to tackle such global challenges and to embrace sustainable living [12–74].

6.4.2 Requirements of a 21st Century Engineer

Engineers with problem-solving, innovating capabilities and good interdisciplinary skills are assets to a nation's economy and their decisions affect local and global communities [77–79]. 21st century engineers are applauded as technological pioneers of modern-day wonders such as artificial intelligence, quantum computing, robotics, automation, geospatial digitalization among many others. But on the other hand, they are also expected to be stewards of the planet leading the advancement of 2030 SDGs [75–82]. They are in a difficult situation since they not only have to work with limited resources but also need to minimize waste generation while increasing the production of goods and services to meet the increasing demands of the growing global population [83]. Engineers have been rising to these challenges by developing methods to produce carbon-neutral forms of energy, designing buildings that use local materials, using components that can be re-used, using alternative sources of energy to provide faster and safer mobility with vehicles. But in the 21st century, engineering is undergoing a profound transformation and is facing complex challenges including those encapsulated in the SDGs requiring multi-disciplinary, cross-country and inter-cultural solutions [72]. Figure 6.6 on page 154 presents an overview of the multitude of skills that 21st century engineers are expected to possess that encompass technical, interpersonal, sustainable development and smart skills all driven by SD principles.

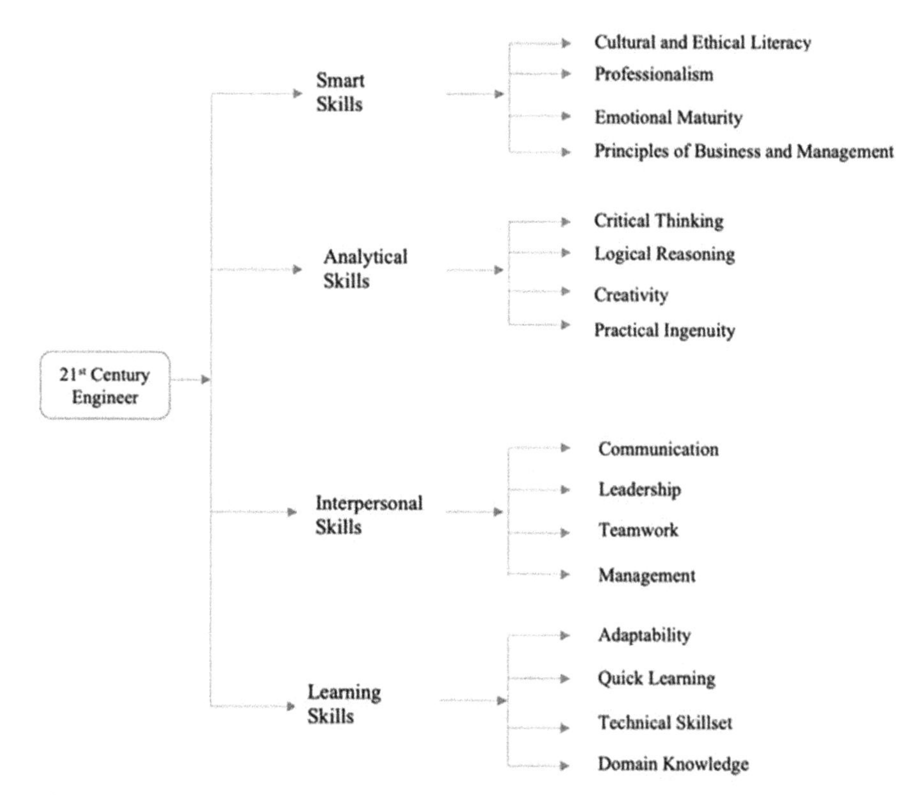

Figure 6.6 Key Skills and Competencies for a 21st Century Engineer [adapted from [75–84]].

6.5 Sustainability Education in the 21st Century

By equipping students with the necessary knowledge and skills for SD, ESD can help them to develop a much better understanding of sustainable entrepreneurial and agricultural practices to be even more eco-actionable [26, 85]. Researchers of ESD recommend place-based, project-based and discovery approaches to teach for sustainability by focusing on local to global impacts of human activities, personal and social sustainability along with promoting collaboration and interdisciplinarity [10, 86]. It is also essential that educational institutions practice what they preach by adopting sustainable practices across the board which would further motivate and lead students towards sustainability. UNESCO recommends decision-makers and members of public and private institutions empower and mobilize youth by recognizing them as key contributors in all SD efforts [12]. More importantly, a holistic approach focusing on skills, perspectives and values should be encouraged by communities and nations to achieve sustainable goals because sustainable societies cannot be formed based on only basic literacy. Figure 6.7 below presents the required skills, perspectives and values that need to be incorporated in the ESD curriculum which is central to SD in each of the environmental, economic and societal aspects of sustainability.

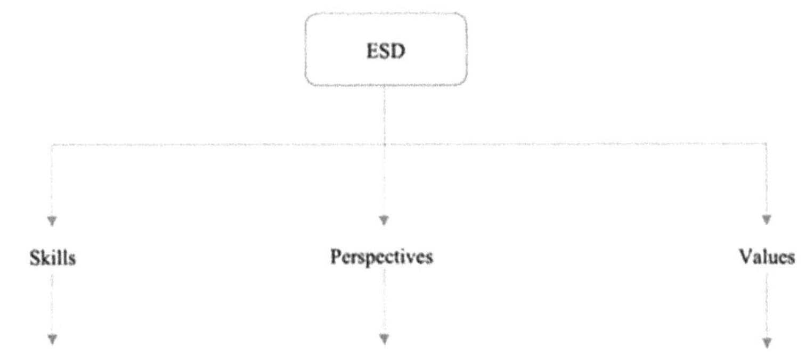

- effective communication
- systems thinking
- foresight to forecast, think and plan ahead
- ability to progress from awareness to knowledge to action
- critical thinking
- awareness about local, cultural, ethical, and environmental issues
- ability to harvest local materials with a focus on reuse, recyclability, and waste minimization

- dynamic nature of societal and environmental problems
- linkages between global issues
- local issues in a global context
- economic, religious, and societal values compete for importance as diverse people interact
- individual decisions and actions impact resource extraction and manufacturing in distant places
- employing precautionary principle by taking action to mitigate serious environmental damage or social harm in the long run

- understanding diverse viewpoints
- community participation
- cultural and social inquisitiveness
- volunteerism and social justice
- respect for diverse traditions and religions
- empathy for life conditions of other people
- ecological sustainability and resource conservation
- preservation of resources for future generations

Figure 6.7 Key Contributions of Education for Sustainable Development [adapted from 87].

Though the 21st century is marked by the rapid adoption of new technologies leading to unprecedented growth and innovation progressing societies and economies to unprecedented heights, there is still much work to be done on the SD front. The pace of SD needs to switch to a higher gear to tackle the novel challenges of the 21st century by incorporating all aspects of sustainability [88]. The 21st century, encompassing the Information Age is fueled by a 'Knowledge Economy' that values problem-solving and critical thinking over the rote skills of the Industrial era. Thus, to remain afloat in the competitive global and knowledge-based economy, ESD must ensure that learners are trained on necessary sustainability skills, values and perceptions to ensure the quality of life for everyone and to facilitate sustenance of our planet [89–93].

6.5.1 Role of Engineers in Sustainable Development

More than ever before, engineers are expected to be able to manage the global nature of the engineering profession with diminishing natural resources coupled with a deeper understanding of the specific local social, cultural, economic and environmental impacts of their work [33]. There is a growing realization and acceptance that sustainable engineering is the route to economic prosperity and an essential tool to tackle most of the biggest challenges that engineering faces in the 21st century [81, 72]. Since sustainability is poised to be the key driver for innovation in the 21st century, engineers should be prepared to contribute bottom-up to a global SD [94, 95]. It is incumbent upon the engineering community to create products and services that are not only economically viable but also environmentally sustainable along with being ethically and socially appropriate [96, 97]. With increasing globalization and improvements in the standard of living worldwide, engineers must continue to innovate in a sustainable manner to address current and future sustainability challenges [98, 97]. Hence, imparting ESD for engineers is of utmost importance so that they are prepared to address the current and impending complex challenges of SD [102–109].

6.5.2 Engineering Education for Sustainable Development (EESD)

The engineering profession has been one of the most dynamic professions that is constantly evolving since the success of engineers and firms greatly depends on their adaptability to new and ever-changing conditions and technologies [89]. Engineers are expected to contribute not only to technical aspects but also to human aspects by pursuing the 'public good' ideal through SD which would benefit our societies, economies, and environment [77]. To meet the complex challenges that engineers face, new domains of 'Sustainable Engineering' and 'Engineering Education for Sustainable Development' (EESD) have been designed following the principles of ESD. Sustainable Engineering is defined as 'the process of using resources in a way that does not compromise the environment or deplete the materials for future generations' [110] whereas EESD can be defined as the education that empowers engineers to innovate in a technically, environmentally and socially responsible manner [111]. Table 6.3 on page 157 gives an overview of the 10 guiding principles for engineers on incorporating SD principles in their work.

Table 6.3 Key Sustainability Competencies and Guiding Principles for Engineers [adapted from 77]

Principle	Guidelines
Maintaining and Improving Knowledge	– possess sustainability knowledge – continuously learn and improve – be familiar with Environmental Management Systems
Limits to Competency	– recognize that sustainability issues require expertise across a range of disciplines – obtain expert advise on environmental issues – undertake only efficiently manageable work
Social Impacts	– identify the effects of proposed actions beyond one's own locality over the long run – ensure that inputs on societal values are considered in engineering and prioritize public good
Sustainability Outcomes	– undertake an environmental assessment process in the early planning stages in compliance with procedures – identify and promote cost-efficient solutions and approaches in integrating environmental, social and economic considerations
Costing & Economics	– conduct an economic analysis of the project – recognize that environmental protection and associated costs are an integral part of project development
Planning & Management	– assess project alternatives to protect and enhance the environment and its sustainability – prioritize the use of local materials, products, and services and adapt to changing conditions

(*Continued*)

Table 6.3 Continued

Principle	Guidelines
Innovation	– innovate solutions that balance environmental, social, and economic factors with limited resources – use knowledge transfer, capacity building and measurement of outcomes to promote innovation
Communication and Consultation	– actions should be guided by accountability, inclusiveness, transparency, commitment and responsiveness – provide clear, timely and complete information
Regulatory & Legal Requirements	– inform public regulatory authorities of all environmental effects of the work – strive to protect the public health andăwell being – prompt action and communication
Risk Mitigation	– carry out a risk assessment, develop actions that address the highest risks and communicate these risks and actions to stakeholders – understand the consequences of specific actionsăandăof inaction

Though some of the competencies and skills presented in Table 6.3 on page 157 might seem far-reaching or irrelevant for certain scenarios or engineering disciplines, these guiding principles could still be used as references to work towards better solutions by being mindful of their overall impact. Long-term thinking and paradigm shift in resource consumption are necessary since sustainability has major implications for engineers [77]. Researchers advocate that the main goal of EESD is to inspire and empower learners with SD thinking [112]. This is critical to meet the growing demand for sustainable engineering and sustainability-skilled graduates. Engineering education needs to better understand, evaluate and optimize the integration of SD principles with traditional curriculum and intensify efforts to train future engineers to address the demands of SD growth.

6.6 Conclusion

Practicing sustainability can no longer only be a matter of compliance with the laws and regulations since we are at a point where if the whole approach towards sustainability is not taken as an essential duty, our societies and planet may reach a point of no return. We no longer have the luxury or privilege to have sustainable practices as a mere formality or put on the backburner but rather it needs to take center stage in all our design and development initiatives. It is incumbent upon the current and future generations to make sustainable living a part of their personal and professional lives. Unsustainable practices by individuals and corporations not only impact them but create a ripple effect that would impact local to global communities. If a sustainability mindset in all our endeavors is not adopted, then the very existence of our planet's fragile eco-systems would be threatened which would ultimately be detrimental to human health and survival.

There is already a 'push from students' with growing concern and willingness to work towards a sustainable future and among businesses to adopt SD principles. Especially, engineers are at the core in the efforts to achieve the 2030 Agenda of SDGs since they can play a significant role and contribute positively to tackle the 21st century challenges related to global sustainability, infrastructure, energy, economic and societal progress. In a fast-paced world, sustainability is critical for industries to succeed. An environmental engineer with a sustainability focus and entrepreneurial mindset can contribute tremendously to the success of the industries' sustainability goals. Industries focused on improving their sustainability impacts seek help from skilled professionals trained in sustainable development. This 'pull from industry' creates opportunities for engineers to utilize their technical acumen and SD knowledge in every aspect of growth and development. Hence, it is essential that educational institutions tap into this rising awareness, interest and a sense of responsibility by helping learners achieve vital skills required for achieving SDGs through ESD.

References

[1] UNWCED. (1987). Our common future, Geneva

[2] United Nations General Assembly (UNGA). (2015). Transforming our world: The 2030 agenda for sustainable development. Available: www.un.org/ga/search/view_doc.asp?symbol=A/RES/70/1&Lang=E [accessed on 06.13.2021]

[3] Rosen, M. A. (2019). Advances in sustainable development research. *European Journal of Sustainable Development Research*, vol. 3, no. 2, em0085. https://doi.org/10.29333/ejosdr/5730

[4] Wastl, J., Porter, S., Draux, H., Fane, B., and Hook, D.W. (2020). Contextualizing SD Research.

[5] Graham, Sarah. (2019). The rise of sustainability degrees and professional positions: Hope for a More Sustainable Future. Available: https://brizomagazine.com/2019/01/17/the-rise-of-sustainability-degrees-

and-professional-positions-a-hope-for-a-more-sustainable-future/. [accessed on 06.13.2021]

[6] International Institute for Sustainable Development (IISD). (2012). Sustainable development timeline. Available: www.iisd.org/system/files?file=publications/sd_timeline_2012.pdf. [accessed on 05.19.2021]

[7] UN News. (2019). The 2010 – 2020 UN News Decade in Review, part two. Available: https://news.un.org/en/story/2019/12/1053701. [accessed on 05.29.2021].

[8] UN Environment Programme (UNEP). (2020). Environmental moments: A UNEP@50 timeline Available: www.unep.org/news-and-stories/story/environmental-moments-un75-timeline. [accessed on 06.19.2021]

[9] Ghorbani, Shervin. (2020). The History of Sustainable Development Goals (SDGs) Available: https://thesustainablemag.com/environment/the-history-of-sustainable-development-goals-sdgs/. [accessed on 06.17.2021]

[10] Moore, J. (2005). Seven recommendations for creating sustainability education at the university level: A guide for change agents, *International Journal of Sustainability in Higher Education*, vol. 6, no. 4, pp. 326–339.ăhttps://doi.org/10.1108/14676370510623829

[11] Wang, Jiawen; Yang, Minghui;ăMaresova, Petra. (2020). "Sustainable development at higher education in China: A Comparative Study of Students' Perception in Public and Private Universities". *Sustainability*. doi: https://doi.org/10.3390/su12062158

[12] UNESCO. (2020). Education for sustainable development. Available: https://en.unesco.org/themes/education-sustainable-development. [accessed on 04.28.2021]

[13] UNESCO. (2021b). UNESCO declares environmental education must be a core curriculum component by 2025. Available: https://en.unesco.org/news/unesco-declares-environmental-education-must-be-core-curriculum-component-2025-0. [accessedon 05.29.2021]

[14] Buchanan, J. (2012). Sustainability education and teacher education: Finding a natural habitat? *Australian Journal of Environmental Education*, vol. 28, no. 02, pp. 108–124. doi:10.1017/aee.2013.4

[15] Tilbury, D. (2004). Environmental education for sustainability: A force for change in higher education. *Higher Education and the Challenge of Sustainability*, pp. 97–112. doi:10.1007/0-306-48515-x_9

[16] Sivapalan, S., Clifford M.J. (2019) Engineering education for sustainable development. In: Leal Filho W. (eds) Encyclopedia of

Sustainability in Higher Education. https://doi.org/10.1007/978-3-030-11352-0_13

[17] York University (YorkU). (n.d). History of ESD - UNESCO Chair in Reorienting Education toward Sustainability. Available: https://unescochair.info.yorku.ca/history-of-esd/. [accessed on 04.26.2021].

[18] Frueh, Sara. (2020). Colleges and Universities Should Strengthen Sustainability Education Programs Available: https://www.nationalacademies.org/news/2020/10/colleges-and-universities-should-strengthen-sustainability-education-programs-by-increasing-interdisciplinarity-fostering-experiential-learning-and-incorporating-diversity-equity-and-inclusion. [accessed on 06.13.2021]

[19] Kerkhoff, L., and Lebel, L. (2006). Linking knowledge and action for sustainable development. *Annual Review of Environment and Resources*, vol. 31, pp. 445–477.

[20] Nature Editorial. (2021). How science can put the Sustainable Development Goals back on track. Available: www.nature.com/articles/d41586-021-00104-0. [accessed on 06.13.2021]

[21] Henderson, K., and Tilbury, D. (2004). Whole-school approaches to sustainability: An international review of sustainable school programs. Report prepared by the Australian Research Institute in

[22] Choudhary, Amit. (2020). Generation Green is leading the sustainability agenda. Available: www.capgemini.com/2020/08/generation-green-is-leading-the-sustainability-agenda/. [accessed on 04.14.2021]

[23] Petro, Greg. (2020). Sustainable Retail: How Gen Z Is Leading the Pack Available: www.forbes.com/sites/gregpetro/2020/01/31/sustainable-retail-how-gen-z-is-leading-the-pack/?sh=586cf9bb2ca3. [accessed on 06.09.2021].

[24] Yamane, Tomomi and Kaneko, Shinji (2021). Is the younger generation a driving force toward achieving the sustainable development goals? Survey experiments. *Journal of Cleaner Production*. vol. 292, pp. 125932 DOI: 10.1016/j.jclepro.2021.125932

[25] Nielsen. (2019). SUSTAINABLE SHOPPERS BUY THE CHANGE THEY WISH TO SEE IN THE WORLD. Available: www.nielsen.com/wp-content/uploads/sites/3/2019/04/global-sustainable-shoppers-report-2018.pdf. [accessed on 05.15.2021].

[26] Spark-Y. (2019). Why Sustainable Education Is Crucial for the Next Generation. Available: www.spark-y.org/blog/2019/11/18/why-sustainable-education-is-crucial-for-the-next-generation. [accessed on 04.20.2021].

[27] Asad, Harun. (2021). Sustainably Minded Customers: The New 'Era of Sustainability'. Available: www.environmentalleader.com/2021/06/sustainably-minded-customers-the-new-era-of-sustainability/. [accessed on 05.10.2021]

[28] Hassim, Alleeya. (2021). Why younger generations are more willing to change in the name of sustainability. Available: www.greenbiz.com/article/why-younger-generations-are-more-willing-change-name-sustainability. [accessed on 04.25.2021].

[29] Guerra, A and Smink, C.K. (2019) Students' perspectives on sustainability. In: Leal Filho W. (eds) *Encyclopedia of Sustainability in Higher Education.* Springer, Cham. https://doi.org/10.1007/978-3-030-11352-0_32

[30] BBC. (2019). Global climate strikes: Millions of children take part in protests to help protect the planet Available: www.bbc.co.uk/newsround/49766020. [accessed on 05.20.2021].

[31] GMA. (2019). How Little Miss Flint is still making a difference in her community, 5 years into water crisis. Available: www.goodmorningamerica.com/living/story/miss-flint-making-difference-community-years-water-crisis-62543899. [accessed on 06.10.2021].

[32] The Nobel Prize. (2014). Malala Yousafzai, The Nobel Peace Prize 2014. Available: www.nobelprize.org/prizes/peace/2014/yousafzai/facts/. [accessed on 06.04.2021]

[33] Bourn, D. (2018) The global engineer. In: Understanding Global Skills for 21st Century Professions. Palgrave Macmillan, Cham. https://doi.org/10.1007/978-3-319-97655-6_9

[34] Best, Elisabeth and Mitchell, Nikita. (2018). Millennials, Gen Z, and the Future of Sustainability. Available: www.bsr.org/en/our-insights/blog-view/millennials-generation-z-future-of-sustainable-business. [accessed on 05.09.2021]

[35] Larsson, Naomi. (2019). 'The youth generation is united': the uni students striking for the climate Available: www.theguardian.com/education/2019/sep/19/campus-is-the-perfect-place-to-disrupt-why-university-students-are. [accessed on 06.02.2021].

[36] Young, Samantha. (2016). 3 ways college students are revolutionizing sustainability. Available: www.greenbiz.com/article/3-ways-college-students-are-revolutionizing-sustainability. [accessed on 06.09.2021]

[37] Teare, Chris. (2016). College Campus Protests Include United Students Against Sweatshops. Available: www.forbes.com/sites/christeare/2016/03/31/college-campus-protests-include-unite

d-students-against-sweatshops/?sh=10fbbf0259a6. [accessed on 06.11.2021]

[38] The Nation. (2014). How Student Activists at Duke Transformed a $6 Billion Endowment Available: www.thenation.com/article/archive/how-student-activists-duke-transformed-6-billion-endowment/. [accessed on 05.19.2021].

[39] Friday, Leslie. (2013). University Approves Gender Neutral Housing. Available: www.bu.edu/bostonia/2013/university-approves-gender-neutral-housing/ [accessed on 06.17.2021].

[40] Powell, Mariah. (2018). MTU students push for sustainability. Available: www.uppermichiganssource.com/content/news/MTU-students-push-for-sustainability-478420853.html. [accessed on 06.14.2021].

[41] Wiles, Katherine. (2018). AS ONE SUSTAINABILITY PLAN NEARS ITS END, STUDENTS PUSH FOR THE NEXT ONE TO BE STRONGER. Available: https://dailytrojan.com/projects/green/sustainability.html. [accessed on 05.29.2021].

[42] Magaoay, Erika. (2020). Students push for sustainability office at BYU. Available: https://universe.byu.edu/2020/02/19/students-push-for-sustainability-office-on-byu-campus/. [accessed on 05.19.2021].

[43] Kessing, Maria. (2021). KTH Students for Sustainability. Available: www.kth.se/blogs/studentblog/2021/05/meet-kth-students-for-sustainability/ [accessed on 06.14.2021].

[44] Kapoor, Tarunika. (2021). Amid COVID-19 pandemic, UC Berkeley organizations push sustainability initiatives. Available: www.dailycal.org/2021/04/22/amid-covid-19-pandemic-uc-berkeley-organizations-push-sustainability-initiatives/. [accessed on 06.18.2021].

[45] D.U. (n.d). USG Sustainability Committee. Available: www.du.edu/sustainability/content/usg-sustainability-committee. [accessed on 06.03.2021]

[46] UCLA. (n.d). Sustainability Action Research. Available: www.ioes.ucla.edu/sar/. [accessed on 05.26.2021]

[47] Collins, Kate I. (2019). Portland students take lead in pushing sustainability in school. Available: www.pressherald.com/2019/02/19/portland-students-take-lead-in-pushing-sustainability-in-school/. [accessed on 06.07.2021].

[48] Newton, Green. (2020). Newton Students & Teachers Push for Sustainability in Schools. Available: https://greennewton.org/read-boston-globe-coverage-of-newton-students-teachers-push-for-sustainability-in-schools/. [accessed on 06.07.2021]

[49] NJSSC. (n.d). The NJ Student Sustainability Coalition. Available: www.njstudentsustainability.com/about. [accessed on 05.11.2021].

[50] Students Organizing for Sustainability (SOS). (2021). Survey of students in higher education around the world. Available: https://sos.earth/wp-content/uploads/2021/02/SOS-International-Sustainability-in-Education-International-Survey-Report_FINAL.pdf. [accessed on 06.13.2021].

[51] Tang, K.H.D. (2018), "Correlation between sustainability education and engineering students' attitudes towards sustainability", *International Journal of Sustainability in Higher Education*, vol. 19, no. 3, pp. 459–472. https://doi.org/10.1108/IJSHE-08-2017-0139

[52] U.S. Bureau of Labor Statistics (2020). Occupational Employment and Wage Statistics Available: www.bls.gov/oes/current/oes172081.htm#nat. [accessed on 05.13.2021]

[53] Our world in data. (2020). Climate change and flying. Available: https://ourworldindata.org/co2-emissions-from-aviation. [accessed on 05.10.2021]

[54] Delta (2020). Delta commits $1 billion to become first carbon neutral airline globally. Available: https://news.delta.com/delta-commits-1-billion-become-first-carbon-neutral-airline-globally [accessed on 06.11.2021]

[55] U.S. Cotton Trust Protocol (2020). Is sustainability in Fashion? Available: https://pages.eiu.com/rs/753-RIQ-438/images/Is%20sustainability%20in%20fashion_Industry%20leaders%20share%20their%20views_FINAL.pdf. [accessed on 06.17.2021]

[56] Nike, (2021). Available: https://purpose.nike.com/2025-targets. [accessed on 06.17.2021]

[57] Bishop, P. L. (2000). "Environmental engineering education in North America." *Water Science and Technology* 41.2

[58] ABET. (2020). Accredited Programs. Available: https://amspub.abet.org/aps/name-search?searchType=program&keyword=environmental%20engineering. [accessed on 06.15.2021].

[59] Harichandran, R. S., Carnasciali, M. I., Erdil, N. O., Li, C. Q., Nocito-Gobel, J., and Daniels, S. B. (2015). Developing entrepreneurial thinking in engineering students by utilizing integrated online modules.

[60] London, J. S., Bekki, J. M., Brunhaver, S. R., Carberry, A. R., and McKenna, A. F. (2018). A Framework for Entrepreneurial Mindsets

and Behaviors in Undergraduate Engineering Students: Operationalizing the Kern Family Foundation's" 3Cs". *Advances in Engineering Education*, 7, no. 1, n1

[61] Gorlewicz, J. L., and Jayaram, S. (2020). Instilling curiosity, connections, and creating value in entrepreneurial minded engineering: Concepts for a course sequence in dynamics and controls. *Entrepreneurship Education and Pedagogy*, vol. 3, no. 1, pp. 60–85.

[62] Huang-Saad, A., Bodnar, C., and Carberry, A. (2020). Examining current practice in engineering entrepreneurship education.

[63] Lu, Mingming. "Integrating sustainability into the introduction of environmental engineering." *Journal of Professional Issues in Engineering Education and Practice* 141.2 (2015): C5014004.

[64] Sababha, B. H., King Abdullah, I. I., Sumaya, P., Al-Qaralleh, E., and Al-Daher, N. (2020). Entrepreneurial mindset in engineering education. *Journal Entrepreneurship Education*, vol. 23, no. 1, pp. 1–14.

[65] Engineering Unleashed (2021). Available: https://engineeringunleashed.com/. [accessed on 06.10.2021]

[66] Alexa, Lidia; Maier, Veronica; ?erban, Anca; Craciunescu, Razvan. 2020. Engineers changing the world: *Education for Sustainability in Romanian Technical Universities*—An Empirical Web- Based Content Analysis. *Sustainabilityă*12, no. 5, pp. 1983. https://doi.org/10.3390/su12051983

[67] SDG. (2020). World Population to Reach 9.9 billion by 2050. Available: https://sdg.iisd.org/news/world-population-to-reach-9-9-billion-by-2050/. [accessed on 04.28.2021].

[68] Dugarova, Esuna and Gulasan, Nergis. (2017). Global Trends: Challenges and Opportunities in the Implementation of the Sustainable Development Goals.

[69] Tuncer, G. (2008). University students' perception on sustainable development: A Case Study from Turkey. *International Research in Geographical and Environmental Education*, vol. 17, no. 3, pp. 212–226. doi: 10.1080/10382040802168297

[70] Cloud, Jamie. (2014). The Essential Elements of Education for Sustainability (EfS): Editorial Introduction from the Guest Editor. Available: www.susted.com/wordpress/content/the-essential-elements-of-education-for-sustainability-efs_2014_05/. [accessed on 06.14.2021].

[71] Pabian, Arnold; Bilińska-Reformat, Katarzyna; Pabian, Barbara. (2021). Future of sustainable management of energy companies in

terms of attitudes and preferences of the younger generation. *Energies*. doi: https://doi.org/10.3390/en14113207

[72] UNESCO. (2021a). Engineering for sustainable development: delivering on the Sustainable Development Goals. ISBN:978-92-3-100437-7 (UNESCO), 978-7-5117-3665-9 (CCTP)

[73] Francis, N. (2017) The advantages and risks of sustainability awareness in the Indian higher education sector. In: Issa T., Isaias P., Issa T. (eds) Sustainability, Green IT and Education Strategies in the Twenty-first Century. *Green Energy and Technology*. Springer, Cham. https://doi.org/10.1007/978-3-319-57070-9_9

[74] Students Organizing for Sustainability (SOS). (2020). Executive summary 2020 survey: "Students, sustainability and education". Available: https://sos.earth/2020-survey-summary/. [accessed on 05.02.2021]

[75] Duderstadt, J. (2010). Engineering for a changing world.

[76] Royal Geographical Society (RGS). (2016). 21st Century Challenges. Available: https://21stcenturychallenges.org/2016/08/03/nexus-thinking/. [accessed on 06.19.2021]

[77] World Federation of Engineering Organizations (WFEO). (2013). WFEO model code of practice for sustainable development and environmental stewardship – interpretive guide. Committee on Eng. and the Environment.

[78] Al-Rawahy, K. H. (2013). Engineering education and sustainable development: The Missing Link. *Procedia - Social and Behavioral Sciences,* vol. 102, pp. 392–401. doi: 10.1016/j.sbspro.2013.10.754

[79] Mohd-Yusof, K., Helmi, S. A., Phang, F. A., and Mohammad, S. (2015). Future directions in engineering education: Ed; *ASEAN Journal of Engineering Education*, vol. 2, no. 1.

[80] Desha, C., Rowe, D., and Hargreaves, D. (2019). A review of progress and opportunities to foster development of sustainability-related competencies in engineering education. *Australasian Journal of Engineering Education,* vol. 24, no. 2, pp. 61–73. doi: 10.1080/22054952.2019.1696652

[81] Wilson, Denise. 2019. "Exploring the intersection between engineering and sustainability education" *Sustainability* vol. 11, no. 11, pp. 3134. https://doi.org/10.3390/su11113134

[82] Qureshi, Asad S and Atif, Nawab. (n.d). THE ROLE OF ENGINEERS IN SUSTAINABLE DEVELOPMENT. *Symposium on Role of Engineers in Economic Development and Policy Formulation.* Available: https://pecongress.org.pk/images/upload/books/

10-311%20Role%20of%20Sustainable%20Dr.pdf. [accessed on 04.29.2021].

[83] Kettering. (2016). The importance of sustainability in engineering Management Available: https://online.kettering.edu/news/2016/07/07/importance-sustainability-engineering-management. [accessed on 05.11.2021]

[84] Padurean, Loredana. (2021). The 10 Smart Skills of the Future Available: www.aacsb.edu/insights/2021/april/the-ten-smart-skills-of-the-future. [accessed on 06.17.2021]

[85] Aina Y.A., Amosa M.K., Orewole M.O. (2019) Students' Perception on Sustainability. *Encyclopedia of Sustainability in Higher Education.* Springer, Cham. https://doi.org/10.1007/978-3-030-11352-0_229

[86] Vanderbilt. (n.d). Teaching Sustainability. Center for Teaching. Available: https://cft.vanderbilt.edu/guides-sub-pages/teaching-sustainability/. [accessed on 06.03.2021]

[87] UNESCO. (2006). Education for Sustainable Development Toolkit. Available: https://en.unesco.org/themes/119915/publications/all. [accessed on 06.11.2021]

[88] Esty, Daniel, C. (2017). RED LIGHTS TO GREEN LIGHTS: FROM 20TH CENTURY ENVIRONMENTAL REGULATION TO 21ST CENTURY SUSTAINABLITY. Available: https://digitalcommons.law.yale.edu/cgi/viewcontent.cgi?article=6186&context=fss_papers. [accessed on 04.14.2021]

[89] Galloway, P. D. (2007). The 21st-century engineer: A proposal for engineering education reform. *Civil Engineering Magazine Archive*, vol. 77, no. 11, pp. 46–104. doi: 10.1061/ciegag.0000147

[90] Stewart, Mark. (2010). Transforming higher education: A practical plan for integrating sustainability education into the student experience. *Journal of Sustainability Education* vol. 1, ISSN: 2151–7452

[91] Aurandt, J. L., and Butler, E. C. (2011). Sustainability education: Approaches for incorporating sustainability into the undergraduate curriculum. *Journal of Professional Issues in Engineering Education and Practice*, vol. 137, no. 2, pp. 102–106. doi: 10.1061/(asce)ei.1943--5541.0000049

[92] Abd-Elwahed, M.S., Al-Bahi, A.M. (2020). Sustainability awareness in engineering curriculum through a proposed teaching and assessment framework. *International Journal of Technology and Design Science.* doi: https://doi.org/10.1007/s10798-020-09567-0

[93] Anholon, R., Rampasso, I.S., Silva, D.A.L., Leal Filho, W. and Quelhas, O.L.G. (2020), The COVID-19 pandemic and the growing need to train engineers aligned to the sustainable development goals. *International Journal of Sustainability in Higher Education*, vol. 21, no. 6, pp. 1269–1275. https://doi.org/10.1108/IJSHE-06-2020-0217.

[94] Pujol, F. A., and Tomas, D. (2020). Introducing sustainability in a robotic engineering degree: A case study. *Sustainability*, vol. 12, no. 14, pp. 5574. doi: 10.3390/su12145574

[95] van Erp, Tim and Kohl, Holger. (2018). Perspectives for international engineering education: Sustainable- oriented and transnational teaching and learning. vol. 21. pp. 10–17. 10.1016/j.promfg.2018.02.089.

[96] Mishra, S. (2010). Engineering curricula in the 21st century: the global scenario and challenges for India.

[97] Davidson, C., Hendrickson, C., Matthews, H.S., Bridges, M.W., Allen, D., Murphy, C., Allenby, B., Crittenden, J., and Austin, S. (2010). Preparing future engineers for challenges of the 21st century: Sustainable engineering. *Journal of Cleaner Production*, vol. 18, pp. 698–701.

[98] Legg, R., Tekippe, M., Athreya, K., and Mina, M. (2005). Solving multidimensional problems through a new perspective: *The Integration of Design for Sustainability and Engineering Education.*

[99] Hanning, A., Abelsson, A., Lundqvist, U., and Svanström, M. (2012). Are we educating engineers for sustainability? Comparison between obtained competences and Swedish industry's needs. *International Journal of Sustainability in Higher Education*, vol. 13, pp. 305–320.

[100] Denton, Denice D. (2010). ENGINEERING EDUCATION FOR THE 21st CENTURY: CHALLENGES AND OPPORTUNITIES. Available: www.nsf.gov/pubs/1998/nsf9892/engineer.htm. [accessed on 04.17.2021].

[101] Accreditation Board for Engineering and Technology (ABET). 2018. Readying today's higher ed students to tackle the world's grand challenges. Available: www.abet.org/wp-content/uploads/2018/11/ABET_Sustainable-Engineering_Issue-Brief.pdf. [accessed on 05.21.2021]

[102] Junyent, M., and deăCiurana, A. M. G. (2008). Education for sustainability in university studies: a model for reorienting the curriculum. *British Educational Research Journal*, vol. 34, no. 6, pp. 763–782

[103] Church, Wendy., Skelton, Laura. (2010). Sustainability Education in K-12 Classrooms. *Journal of Sustainability Education*. vol. 1, ISSN, pp. 2151–7452

[104] Watson, M. K., Noyes, C., and Rodgers, M. O. (2013). Student perceptions of sustainability education in civil and environmental engineering at the Georgia institute of technology. *Journal of Professional Issues in Engineering Education and Practice*, vol. 139, no. 3, pp. 235–243. doi: 10.1061/(asce)ei.1943--5541.0000156

[105] Wiek, A., Bernstein, M., Laubichler, M., Caniglia, G., Minteer, B. and Lang, D. (2013). A global classroom for international sustainability education. *Creative Education*, vol. 4, pp. 19–28. doi: 10.4236/ce.201 3.44A004

[106] Remington-Doucette, S.M., Hiller Connell, K.Y., Armstrong, C.M. and Musgrove, S.L. (2013). Assessing sustainability education in a transdisciplinary undergraduate course focused on real- world problem solving: A case for disciplinary grounding. *International Journal of Sustainability in Higher Education*, vol. 14, no. 4, pp. 404–433. doi: https://doi.org/10.1108/IJSHE-01-2012-0001

[107] Olsen, S. I.,ăFantke, P., Laurent, A., Birkved, M., Bey, N., and Hauschild, M. Z. (2018). Sustainability and LCA in Engineering Education – A Course Curriculum. Procedia CIRP, vol. 69, pp. 627–632. doi: 10.1016/j.procir.2017.11.114

[108] Malik, M. N., Khan, H. H.,ăChofreh, A. G., Goni, F. A.,ăKlemes, J. J., and Alotaibi, Y. (2019). Investigating students' sustainability awareness and the curriculum of technology education in Pakistan. *Sustainability*, vol. 11, no. 9, pp. 2651. doi: 10.3390/su11092651

[109] Redman, A., Wiek, A. and Barth, M. (2021). Current practice of assessing students' sustainability competencies: a review of tools. *Sustainability Science*. https://doi.org/10.1007/s11625-020-00855-1

[110] UNESCO. (2017). Sustainable Engineering. Available: www.unesco.o rg/new/en/natural-sciences/science-technology/engineering/sustainabl e-engineering/. [accessed on 06.17.2021]

[111] Byrne E, Desha C, Fitzpatrick J, Hargroves K (2010) Engineering education for sustainable development: a review of international progress. In: 3rd international symposium for engineering education. University College York, Cork, pp. 1–42.

[112] Kastenhofer K, Lansu A, van Dam-Mieras R, Sotoudeh M (2010) The contribution of university curricula to EESD. Gaia vol. 19, no. 1, pp. 44–51.

7

Unleashing Emotions: The Role of Emotional Intelligence Among Students in Upholding Sustainable Development Goals

Christabel Odame[1,*], Mrinalini Pandey[1], and David Boohene[2]

[1]Department of Management Studies, Indian Institute of Technology-Dhanbad, India
[2]School of Management Sciences & Law, University of Energy and Natural Resources, Ghana
E-mail: codame@anuc.edu.gh; mrinalini@iitism.ac.in; dboohene@gmail.com
*Corresponding Author

Abstract

Students ability to recognize their emotions and that of others are very essential. As a result, educating students to be emotionally intelligent helps them to manage their emotions, learn strategies to interact with the environment, deal with critical situations such as loss of a loved one, financial mismanagement or shocks and being a good citizen in general. Emotional intelligence (EI) serves as a tool for handling both positive and negative emotions and this could help in achieving sustainable development goals (S.D.G). The essence of this study is to examine the role of emotional intelligence among students in promoting sustainable development goals.

Secondary data was used for this study. Data was gathered from previous literature and scholarly articles relating to EI and sustainable development goals. A conceptual model is proposed linking the importance of EI in sustainable development goals.

The findings of the study indicated that EI is important in sustainable development goals. Again, the study inspires the interest in developing the emotional intelligence of students not only for the short term but also for the long term.

In the nutshell, higher educational institutions can consider EI as one of the indispensable skills needed in developing sustainable goals.

Keywords: Sustainable Development Goals (SDG), Emotional Intelligence (EI), Students, Higher Education

7.1 Introduction

In this day and age, educational institutions should not only focus on improving the skills of students to help them maximize academic performance rather it should include the soft skills of being able to cooperate and being compassionate in society [3]. By doing this, students tend to understand themselves and others so as to make relating with others much easier. This does not only look out for the short-term relationship with peers but the long-term sustainability of a relationship. Emotional intelligence refers to the ability to identify and manage one's own emotions, as well as the emotions of others [4]. This ability would help students build and maintain relationships, handle finances, as well as environmental issues. Thus, achieving sustainable development goals suggests concern for environmental issues, finances and social issues.

For this reason, educating students to be aware of their emotions and to know the role emotional intelligence plays in upholding sustainable development goals implies helping them solve emotional-related issues, to learn better strategies to deal with them. Also, they will be able to efficiently relate to the environment as a whole. Again, they will be able to deal with disrupting emotions that might emanate from handling finances and deal with themselves and others better.

Previous studies show that children who are given emotional education are able to remain calm and composed in critical situations [6]. They are able to handle negative emotions much easier. Also, results from previous research suggest that educating students to be emotionally intelligent improves the students social-emotional skills, general attitude about their selves and others, academic performance and positive behaviour as negative behaviour such as bullying, aggression and dropout rate reduces [9].

With all this said, this paper examines the role of emotional intelligence in the sustainable development goals of students in higher education. Secondary data is used for this study. Data is gathered from previous literature and scholarly articles relating to EI and sustainable development goals. A conceptual model is proposed linking the importance of E.I in sustainable development goals. The study commences with a review of literature, methodology, conceptual framework, discussion and ends with conclusions and recommendation for future research works.

7.2 Review of Literature

7.2.1 Higher Education and Sustainable Development Goals

Possibly education seems like the only avenue where communities get to develop the inner self of the younger generation who are spearheading the technological age for the future [8]. In years past, the basic aim for education was to inculcate life-preserving skills to the learner nonetheless, in recent times education brings out particular desired changes in learners. That is learners are enabled to live a more determined and important life with emotional, spiritual and analytical qualities.

Thus, to prepare learners for a satisfying life both professionally and personally, the role of education and educators can't be left out. Education should be such that learners are ready to receive and give back to society in a beneficial way. To achieve this, learners should possess an emotional quotient, intelligent quotient and spiritual quotient [8]. These skills will help learners live satisfying in both the short and long term.

7.2.2 Emotional Intelligence and Sustainable Development Goals

Lately, there has been a revolution of emotions and sustainability through emotional analytics and emotional culture [5]. Therefore, sustainability through emotional intelligence abilities is the basic plan for sustainable development goals. Thus, the right mix of education should include both intelligent and emotional quotients to help learners achieve sustainability. As intelligent quotient will help in developing analytical skills of the learners to deal rationally, the emotional quotient will help them to be empathetic and emotional balance as they deal with the environment, finances and the society entirely. The competencies of emotional intelligence such as self-awareness, self-management, motivation, empathy and social skills would enable learners to attain sustainable development goals.

7.2.3 Emotional Intelligence and Environmental Issues

To sustain the environment and to ensure that future generations can live and survive on the planet in the long-term, humans need to be friends of the planet and engage in friendly practices that will sustain the ecosystem. Knowing the environmental concerns of the planet, such as global warming, greenhouse effect, pollution, etc it is reasonable to consider environmentally friendly practices and behavior to sustain it.

Environmental sustainability can be defined as, 'the urgent need to use the earth's resources in ways that will allow human beings and other species to continue to exist acceptably on earth in the future' [10]. The definition places emphasize on humans to act acceptably in regards to sustaining the environment. This connotes that, the way individuals live can affect the environment. Researchers opine that most of the critical environmental problems are a result of human behavior as oppose to issues of science [12]. [7] argued that 'environmental problems are behavioral problems caused by thoughts, beliefs and values that guide human behavior' [11, 13].

In psychology, human behavior is studied to understand how people relate to their natural surroundings either negatively or positively. More so, to find more beneficial ways to relate with the surroundings, it is prudent to know the role of individuals as well as alter their behavioral patterns to be more environmentally conscious. Emotional intelligence can help individuals to know their and others' emotions so that they can act rightly.

7.2.4 Emotional Intelligence and Society or Social Issues

Self-awareness and 'others aware' really help in building and maintaining relationships. Whether relating to yourself or with people like colleagues or family members, this awareness will help one to know how their emotions affect those around them and inform them of some behavioral patterns. On the other hand, knowing the emotions of one's self and others helps in dealing with it.

Previous studies [1, 2] suggest that the social setting one finds themselves in can affect the development of higher emotional intelligence with time. For instance, living with a family with high parental warmth, better family relationships, etc are all associated with higher emotional intelligence.

7.2.5 Emotional Intelligence and Economy or Financial Issues

Emotions are complicated and very informative so being in tune with them could help make a meaningful financial decisions that would result in greater fulfillment. Identifying and understanding emotions will help prevent hasty financial decisions especially in times of crisis. When conditions are not conducive and favourable, it is very easy to react with anger or fear which might lead to the selling off all your stocks for example during a global pandemic. In the phase of confusion and frustration, one might easily ignore certain signals in relation to finances. Being aware of your emotions and that of others will be of great help to bring sustainability even in difficult financial situations. Thus, leveraging your emotional intelligence can create a space for your emotions for you to understand them rather than allowing your mood to drive your actions. This will help you to be strategic as possible in regard to making financial decisions.

Recent literature identifies the essence of emotions in the economy because through emotional intelligence or emotional management one is able to better perform economic roles and this is often because it aids in logical thinking. However, most of the current research focuses on the individual points of view.

7.2.6 Methodology

Basically, works of previous scholars and popular literature were reviewed. That is, secondary data was used. The keyword's emotional intelligence and sustainable development goals were the target to be explored to know the role of Emotional intelligence in sustainable development goals. Both empirical and qualitative studies were explored.

An independent variable (emotional intelligence) and a dependent variable (sustainable development goals) were identified. Therefore, this study purports the importance of emotional intelligence in sustainable development goals.

7.3 Results

A model is proposed, where the importance of emotional intelligence in sustainable development goals is revealed. Thus, how emotional intelligence helps to attain sustainability considering the three core pillars of sustainable development goals; environment, society and finances.

Proposed Conceptual Model for Emotional Intelligence and Sustainable Development Goals.

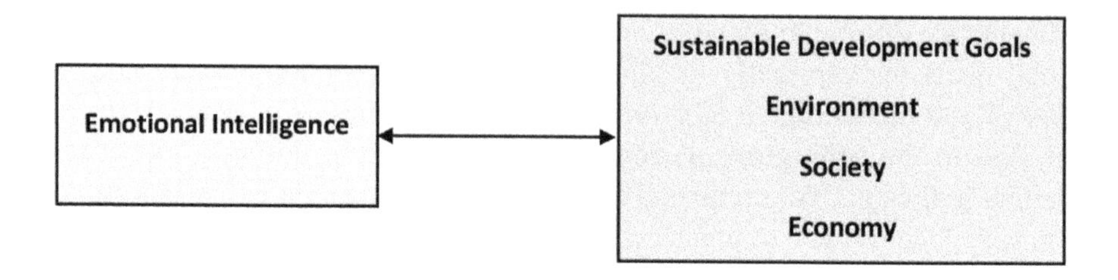

7.4 Discussion

Emotional intelligence constructs (as per the mixed model of Daniel Goleman), must be considered while educators are mentoring or teaching learners in the higher education institution, as such soft skills will enable these learners to know about themselves and the people in their surroundings. This knowledge will guide learners to relate better in the now and in the future while lingering on for the sustainability of other events. Considering what EI is, and from the works of previous scholars, it can be said that EI will be of importance while sustainability is sought after.

First of all, the importance of EI in environmental issues is glaring. As in, to have sustainability humans needs to be more friendly to the ecosystem, that is the behavior of people, could greatly affect the environment. This balls down to how individuals behave. As such when one is emotionally intelligent positive emotions such as love, care, a tender can be channelled to make the environment more beautiful and pleasant to occupy. While negative emotions can be curbed so that they will not have any effect on the environment. As more and more students in higher institutions become aware of their emotions and they are emotionally intelligent, how they behave would change having a tremendous effect on the environment. For example, looking at the forestry, more trees will be planted and the greenery scenes will also be the natural habitat for the animals as there would be peaceful coexistence. On the other hand, with ocean and water bodies, chemicals and other poisonous toxins will not be put into it and consequently killing the fishes and other aquatic animals. Also, freshwater will be available for consumption for a much longer period of time. Correspondingly, pollution, in general, will be decreased, as environmentally friendly practices will be practiced. With all these benefits, EI had a role to play ensuring individuals in this case students behave well so as not to destroy the environment with its natural habitats for future generations to also enjoy it.

Secondly, being emotionally intelligent helps to relate better to yourself and then towards others. In effect, when students become emotionally intelligent dealing with other colleagues, teachers, friends and even people from society would be much easier. With knowledge from EI, they would know better; knowing when to delay gratification, stifle impulses to prevent any unwarranted mishaps from happening. Living well with people around could cause a long-lasting relationship to be built. Legacies can be established for future generations to enjoy. When there is peaceful cooperation amongst students, educators, etc., there wouldn't be strife and malice. Another thing is that good relationships could be built and established. Keeping in mind that, is not just for the now but for the future as well. Proper communication should be established, letting the other party know how you might be feeling or how they made you feel so that you can hold the bull by its horn and properly manage any distressing situations.

Lastly, emotional intelligence and finances, students would have to recognize this so that it serves as a guide as they deal with finances day in and out. Emotions form a great deal of life and it informs us of what to do and not to do. Finances in themselves come with risk due to this it requires sound judgment while dealing with such issues. Being emotionally intelligent will help you harness emotions that will slide into your way so that you can be as objective as possible while dealing with financial decisions. Also, how to handle loss and misfortune will be much better when one is emotionally intelligent and again, how to use money efficiently.

From the aforementioned, it can be realized that emotional intelligence would be much beneficial to students in a higher institutions which will help in achieving sustainable development goals.

7.5 Conclusion and Recommendation for Future Research

In the nutshell, EI can be consider as one of the indispensable skills needed in developing sustainable goals. This is because emotionally intelligent students will be aware of their emotions and consequently managee them so as to achieve sustainability. This skill will help them both in the now and in the future. Future research works can focus on how emotional intelligence will be of help in other industries or sectors as the current study primarily looked at students in higher education.

References

[1] Brackett, M. A., Mayer, J. D., and Warner, R. M. (2004). Emotional intelligence and its relation to everyday behaviour. *Personality and Individual Differences*, vol. 36, no. 6, pp. 1387–1402.

[2] Ciarrochi, J., Chan, A. Y., and Bajgar, J. (2001). Measuring emotional intelligence in adolescents. *Personality and Individual Differences*, vol. 31, no. 7, pp. 1105–1119.

[3] Estrada, M., Monferrer, D., Rodríguez, A., and Moliner, M. Á. (2021). Does emotional intelligence influence academic performance? The role of compassion and engagement in education for sustainable development. *Sustainability*, vol. 13, no. 4, pp. 1721.

[4] Goleman D. P. (1995), Emotional intelligence: Why it can matter more than IQ for character, health and lifelong achievement, Bantam, New York.

[5] Han, D., Park, H., and Rhee, S. Y. (2021). The role of regulatory focus and emotion recognition bias in cross-cultural negotiation. *Sustainability*, vol. 13, no. 5, pp. 2659.

[6] Idoiaga, N., Berasategi, N., Eiguren, A., and Picaza, M. (2020). Exploring children's social and emotional representations of the Covid-19 pandemic. *Frontiers in Psychology*, vol. 11, pp. 1952.

[7] Koger, S. M., and Winter, D. D. (2011). The psychology of environmental problems: *Psychology for Sustainability*.

[8] Kumar, V. V., and Tankha, G. (2021). Nurturing spiritual sntelligence in the classroom. In T. Madden-Dent, and D. Oliver (Ed.), *Leading Schools With Social, Emotional, and Academic Development (SEAD)* (pp. 187–201). IGI Global. http://doi:10.4018/978-1-7998-6728-9.ch010

[9] Martínez-Martínez, A. M., López-Liria, R., Aguilar-Parra, J. M., Trigueros, R., Morales-Gázquez, M. J., and Rocamora-Pérez, P. (2020). Relationship between emotional intelligence, cybervictimization, and academic performance in secondary school students. *International Journal of Environmental Research and Public Health*, vol. 17, no. 21, pp. 7717.

[10] Oskamp, S. (2000). A sustainable future for humanity? How can psychology help?. *American Psychologist*, vol. 55, no. 5, pp. 496.

[11] Smith, J., Shearman, D., and Positano, S. (2007). Climate change as a crisis in world civilization: why we must totally transform how we live. *Edwin Mellen Press*.

[12] Voulvoulis, N., and Burgman, M. A. (2019). The contrasting roles of science and technology in environmental challenges. *Critical Reviews in Environmental Science and Technology*, vol. 49, no. 12, pp. 1079–1106.

[13] Winter, M. (1996). Rural politics: policies for agriculture, forestry and the environment. *Psychology Press*.

8

Pedagogy for Living in Harmony with Nature – Sustainability in Higher Education

Qudsia Kalsoom[1], and Sibte Hasan[2,*]

[1]Beaconhouse National University, Pakistan
[2]Independent Researcher, Pakistan
E-mail: qudsia.kalsoom@bnu.edu.pk
*Corresponding Author

Abstract

Human beings have lived as a part of Nature for millions of years. In these years, humans' relationship with Nature remained harmonious. Earlier humans believed that the objects of Nature (living and non-living) had souls and feelings. With the agricultural revolution, humans started domesticating animals and plants. However, their relationships with Nature did not change much. The rise of capitalism in the past three centuries has not only changed human relationship with Nature but also with other human beings. Now Nature is 'other' to human beings. On one hand, this 'otherness' has allowed humans to exploit Nature's resources and on the other hand, it has replaced human relationships with commodities leading to promoting consumer culture and unsustainable development. Higher education is largely viewed as a key enabler for sustainable development. Over 500 universities across the globe have signed the Talloires Declaration, a ten-point action plan for incorporating environmental literacy and sustainability in research, teaching and operations. Sustainability teaching in higher education may aim at developing students' capacities as critical beings who do not follow dominant consumer ideology or cultural and environmental practices rather question them. They should be able to look at the issues of sustainability in the larger historical context. To achieve the desired outcomes, the pedagogy may include transdisciplinary learning content (the content related to United Nations Sustainable Development Goals) and the processes of dialogue, collaboration and critical reflection. The underlying assumption for this pedagogy is that unsustainability is a social issue and it may be addressed if students' capacities are developed as critical beings.

Keywords: Higher Education, Sustainable Development, Pedagogy of Sustainability, Critical Thinking Dispositions, Transformative Learning.

8.1 Introduction

> We are the land . . . that is the fundamental idea embedded in Native American life: the Earth is the mind of the people as we are the mind of the earth. The land is not really the place (separate from ourselves) where we act out the drama of our isolate destinies. It is not a means of survival, a setting for our affairs . . . It is rather a part of our being, dynamic, significant, real. It is our self . . . It is not a matter of being 'close to nature' . . . The Earth is, in a very real sense, the same as our self (or selves). Paula Gunn Allen [1].

Traditionally universities have been recognised as sites for the creation of scientific knowledge for industrial innovation and preparation of skilled labour force [2]. Being knowledge creator spaces, they have served as ideological apparatuses as well [2]. States have used institutions of higher education to promote national identity through the development of historical and cultural narratives [3]. However, the role of universities is beyond producing a labour force and creating knowledge for industry [4]. Universities need to play a role in social transformation and development. Literature indicates that universities significantly affect people's capacities and dispositions as citizens [3]. Scholars [5–7] from the field of higher education for sustainable development (HESD) have also emphasized upon the transformative role of universities. They believe that universities need to perform a lead role in achieving sustainability. Heleta and Moodien [8] also argue that SD cannot be achieved anywhere in the world 'without the capacity-building contribution of an innovative higher education system' because higher education is directly linked to many SDGs such as SDG1, poverty reduction; SDG3, health and well-being; SDG5, gender equality; SDG 12, responsible consumption and production and SDG 13, climate change [9]. Purcell et al. [10] declared that universities are engines to achieve SDGs. They further argue that higher education can make a fuller contribution to sustaining the environmental, economic, cultural and intellectual well-being of global communities by building partnerships within and with universities.

Realizing the critical role of universities in advancing the agenda of SD, universities have taken different initiatives. Wu et al., [11] in their survey of 642 business schools noted that many universities had included sustainability courses in different academic programmes. As a response to UN Agenda 21 (for ESD), universities from different countries have signed the Talloires Declaration to promote HESD at political and policy levels [12]. Some of the declarations and programmes are: Johannesburg Declaration [13]; UNECE Strategy on ESD [14]; and Bonn Declaration in 2009. Recently, Times Higher Education (THE) has started ranking universities on the basis of their contribution towards sustainable development goals (SDGs). To contribute more effectively towards SDGs, universities need to make SDGs an integral part of their overall mission. Hall and Tandon [15] also suggest to innovate higher education missions. They contend that the missions should be moulded in a way that helps achieve SDGs in the long run.

Sustainability education is a key enabler towards achieving all other SDGs [16]. Therefore, it is important to unpack the concept of sustainability education especially with reference to higher education. Mostly, HESD has been described with reference to its outcomes or targets such as sustainability competencies, sustainability capabilities, sustainability literacy, sustainability learning, sustainability consciousness [17–19]. These targets or outcomes provide a reference to define and explain HESD. For example, with reference to sustainability competencies, HESD can be understood as an education that ultimately leads to the development of sustainability competencies. Similarly, with reference to sustainability consciousness, HESD may be defined as higher education that leads to enhancing awareness about the issues of sustainability and leading to the development of pro-sustainability behaviours and attitudes. However, some scholars [20, 21] argue that sustainability education cannot be close-ended because the sustainability problems (social, economic and environmental) are not definite. Jickling and Wals [22] find identifying the exact end-states as mis-educative. They argue that sustainability education should aim at developing students' capacity to foresee future problems and address them. The pandemic situation (Covid-19) also supports the argument that the future is unpredictable and uncertain. Gill [23] explains the unpredictability of the future by highlighting that human activities are helping diseases to spread from animals into humans more frequently.

The impact and scale of these diseases cannot be predicted at this stage. Keeping in view the unpredictability of the future, education should aim at developing people's capacities as critical beings so that they could respond to unforeseen situations in the best possible ways. In other words, sustainability education should transform students' thinking or frames of reference.

The transformative role of education and higher education for sustainability has been discussed in the literature [24, 25]. Sterling maintains that the concept of sustainability education resonates with Freire's concept of 'critical consciousness' and Mezirow's concept of transformative learning. He further argues that sustainability education should aim at transforming students' worldviews [24]. Kalsoom and Shah [25] describe the transformation in students' worldviews as epistemological development. They argue that sustainability education and teaching should essentially aim at transforming the epistemological beliefs of the students. In a recent study, Shephard et al. [26] found a correlation between critical thinking dispositions (open-minded and fair-minded thinking) and the ecological worldview of people. They argue that thinking dispositions may be indicators of learning gains towards indicative HESD learning targets such as sustainability concerns [26]. Building on the argument of Shephard et al., it may be assumed that development of critical thinking dispositions should be the focus of HESD since it directly affects people's sustainability concerns.

The critical thinking dispositions may be understood in terms of Mezirow's [27] concept of perspective transformation. According to him, our perspective or frames of reference is a set of assumptions that direct the interpretation of our experiences [27]. They could be inclusive or exclusive; open or narrow. Taylor [28] believes that meaning perspectives 'are like a 'double-edged sword' whereby they give meaning (validation) to our experiences, but at the same time skew our reality'. So it is important to question one's meaning perspectives and change those which are not reflective, inclusive or open. Such frames of reference are *'problematic'* because they could be 'detrimental to personal development and growth' [29]. The process of changing the *problematic* or *detrimental* frames of reference is a perspective transformation or transformative learning. Kegan [30] describes the epistemological transformation is shifting 'away from being 'made up by' the values and expectations of one's 'surround' (family, friends, community, culture)...toward developing an internal authority that makes choices about these external values and expectations according to one's own self-authored belief system' (p. 46).

The current chapter aims at identifying the problematic ecological frames of references with respect to our patterns of production and consumption and proposes a pedagogy that can help in transforming those frames of reference. The first section of the chapter (Capitalism and Human Behaviour) highlights the capitalist practices of production and how they influence people's meaning perspectives regarding the consumption of materials. This is achieved through the pedagogy of consumerism as it promotes ahistorical relation with Nature. In ahistorical relationship, ideas are developed without their historical contexts. Practices are focused on goals to be obtained for immediate advantages and are value-free. The second section of the chapter proposes and discusses the 'pedagogy of harmony with Nature' (a transformative pedagogy). The underlying assumption of the current chapter is that sustainability is a social issue and can be addressed by employing a pedagogy that can bring changes in worldviews or lead to epistemological development.

8.2 Capitalism and Human Behaviour

The rise of capitalism led to redefining the human relationships with Nature and other human beings. Earlier humans viewed themselves as a part of Nature. They used Nature's resources to fulfill their existential needs. However, the industrial revolution and the rise of capitalism separated human beings from Nature. Humans started viewing Nature as 'other'. As a result of this 'otherness', humans have universalized the exploitation of Nature's resources. In the early days of capitalism, the whole world was colonized but it remained rooted in nation states. After the mid-twentieth century, the capitalist system has been transformed into an empire. An 'empire establishes no territorial centre of power and does not rely on fixed boundaries or barriers. It is a de-centred and deterritorializing apparatus of rule that progressively incorporates the entire global realm within its open, expanding frontiers' [31]. The empire of capitalism has influenced the whole planet to such an extent that scholars (Crutzen and Steffen) have labeled this age as 'Anthropocene' or the age of mankind [32, 33]. Humans are controlling the production of millions of things such as the creation of spaceships, creating germs and even the evolution of species. Capitalism is the driving force behind the 'anthropocene'.

The capitalist system does not restrict itself to the production of commodities and accumulation of capital; it affects society as a whole. Negri and Hardt [31] argue that the capitalist system tends toward biopolitical production (the production of social life itself) in which the political, economic and cultural systems overlap and invest in one another. In other words, this system is geared to 'produce needs, social relations, bodies and minds —. In the biopolitical sphere, life is made to work for production and production is made to work for life' [31].

To achieve its biopolitical goals, the capitalist system has evolved an ideological superstructure of consumerism. 'Capitalist production transforms the relations of individuals into qualities of things themselves and this transformation constitutes the nature of the commodity in capitalist production' [34]. Marx labeled it as 'the fetishism of commodities'. In another work, Fromm argues that 'modern consumers may identify themselves by the formula: I am = what I have and what I consume' [35]. In consumerism, human relations are not only disguised behind objects but also are replaced by them. When a commodity loses its use-value, a symbolic character is superimposed on it. A fetish possesses intrinsic awe and magic. Objects are basically tools to make life easier and comfortable. They have use-value. However, if these are possessed for possessing sake, the driving force is consumer ideology.

In the early days of consumerism, as claimed by Fromm [35], the 'having' mode of life was propagated and lived. The 'having' mode referred to having the things to gain inner satisfaction. Later, the having mode transformed into the 'wanting' mode. In this mode, a consumer is encouraged to develop a psyche of a collector. Collectors are not interested in the use-value of an object. They always have a sense of emptiness and that has to be filled by getting the desired objects at any cost. And this craving does not end after the possession of an object. The desire of a consumer always exceeds the necessity. The limitlessness of the desire becomes an endless longing for the new. This is the essence of the ideology of consumerism. In this ideology, 'appearing' or spectacle is important. Lefebvre and Nicholson-Smith have explained the ideology of consumerism by referring to the ideas of Nietzsche who had pointed out how over the course of history, the visual has increasingly taken precedence over elements of thought and action deriving from the other senses [36].

8.2.1 Capitalism and Communication Industries

The communication industries convince people that every salable product is necessary and they need this anyway. This industry says that this object is desirable and therefore people should desire it. If one does not follow this mantra, one would be missing something very important. These suggestions work silently and buying and displaying things becomes an emotional impulse. This strong drive of consumption is always justified by the consumer in some way [37]. If objects are possessed for possessing sake, it means the consumer's act is socially driven or has roots in one's psyche which is manifested in this act of possessing some objects. In such a case, the purchased object creates a spectacle of buying. In this case, buying becomes a 'sign production' [37]. A sign represents something other than itself. For example, traffic signs represent guidance for commuting on roads, not the boards themselves. Signs become alive when used. Similarly, in consumerism, commodities become signs and they represent feelings of 'wanting' and then an impulse to the spectacle. These signs remain mute unless their presence is acknowledged. In this situation, the interaction with the signs of commodities becomes metaphorical. 'The metaphorized thirst transforms both the objects of consumption and the objects of desire into a chain of substitute, substituting one thing for another and thus offering alternative objects for the fundamentally 'objectless' desire to be fixed upon' [37].

The ideology of consumerism is virtual. It is a system in which people's material symbolic existence is entirely captured. It is fully immersed in a virtual image setting, in the world of make-believe. Castells [38] argues that this is not a fantasy rather a material force because 'it informs, and enforces powerful economic decisions at every moment in the life of the network' [38]. In such a situation we see a reality which is produced by the commodity signs.

Consumerism has a concrete cause and effect chain behind its intentions and operations. That is why its strategies change with the change of historical settings. For example, as mentioned earlier, in the beginning, it provoked people to possess objects and with the emergence of information and the age of networking it promoted a sense of 'wanting'. In this stage of 'wanting' a consumer feels an insurmountable sense of lacking. Consumerism is very tactful and pragmatic. It works by promoting irrationality and abstraction. In an anthropological sense, consumerism has got a status of a religion. Religion or abstractions based upon religious/mythic ideas are the means to unite people under one banner. When an emperor embraces a religion, he/she does not become pious or faithful. He/she does this out of necessity or as part of his/her political strategy to unite people under his/her. So an individual is not the architect of an ideological superstructure but a captive in the representations of this ideology. Consumerism is like state religion or state ideology. This has to be reinforced by taking various measures so that it is kept alive in the minds of the people.

8.2.2 Plunder of Nature

Pedagogy of consumerism has not only transformed humankind but also manipulated Nature to maximize the production of consumable objects. During the last two centuries, the Natural resources have been plundered to the extent that now the resources available from one earth are not sufficient. Global Footprint Network reports a steady increase in the ecological footprints of most of the countries since 1961. According to this report, the ecological footprint of the people living in rich countries is very high. The lifestyles of the people of rich countries require more than three earths to live. For example, Qatar needs almost nine earths; UAE requires 5.75 earths; Luxemburg requires nearly eight earths; North America requires 5.1 earths; Europe requires three earths; and Oceania needs 4.5 earths. Low income or middle-income countries require one or less than one earth. The world today needs 1.7 earths to fulfill the needs of all the people. However, we have only one earth.

Excessive production and consumption of commodities are necessary for the success of the capitalist economy. Excessive production needs excessive use of raw materials. Humans have mostly taken raw materials from the earth or natural resources. As a result, natural resources have depleted drastically. This has led to scarcity of freshwater, forests, non-renewable energy resources, sand, etc. Beiser [39] noted that people use nearly 50 billion tonnes of 'aggregate' (sand and gravel) every year and this amount is more than enough to blanket the entire United Kingdom. In addition to depletion in natural resources, more production of things leads to creating more waste in the form of harmful gases, fluids or solids.

World Resources Institute reports that the global emissions of carbon dioxide were approximately 2.5 billion tons in 1,900. However, they doubled in the next 50 years and raised to nearly 5.5 billion tons. In the next 50 years, they reached 24 billion tons [40]. Similarly, plastic waste has emerged as one of the major environmental issues. Plastic dumped in oceans is estimated to kill millions of marine animals every year. Nearly 700 species, including endangered ones, are known to have been affected by it [41]. In 2017, the earth had 6.9 billion tons of waste plastic and out of this 6.9 billion tons, 6.3 billion tons of plastic never made it to a recycling bin. The issue of plastic trash has become so ubiquitous that it has prompted efforts to write a global treaty negotiated by the United Nations [42].

Another addition to unnecessary waste production is online shopping. Customers order things online and if on receiving, the thing does not match their choice, these are returned back. The shipping companies throw them in godowns. This trend has tremendously increased the number of returned items which are generating more than 2.27 billion Kg of waste every year [43].

Massive urbanization, monstrous industrialization, exponential increase in synthetic chemicals, risks of nuclear wars, pandemics, climate change and destruction of biodiversity have led scientists around the world to ask the question 'will humans survive the century?' The Centre for the Study of Existential Risk at Cambridge University brings together scientists from different parts of the world to study the impacts of human activities. The current situation of earth necessitates re-thinking human relationship with Nature.

8.2.3 Pedagogy of Domination/ Consumerism

Domination is the bedrock of consumerism. People silently surrender before consumerism. Hence, consumerism operates as an ideology of oppression like any other totalitarian ideology like Fascism or Stalinism. It operates through a pedagogy which is essentially a pedagogy of domination. The pedagogy of consumerism breeds a cognition that automatically starts dominating not only human beings but also uses human beings as the tools of domination. In this pedagogy, apparently, people buy commodities and assume that they are using their agency, but in reality, the communication industries manipulate them to buy an object. Buying an object is portrayed as desirable and necessary. This manipulation turns into a collective sentiment that has to be imitated and hence followed to remain a part of the crowd. In a crowd, an individual though surrenders its subjectivity but feels protected anyway. So, a 'reality' is produced which is not based on facts but on simulacrum. In this situation, people become a pathway on which representations of commodities flow but they themselves have no control over it. This is the objectification of human beings.

The process of objectification starts under the hegemony of the dominating groups. This is equally repressive and violent as of the violence perpetrated in the pre-modern eras. In the pre-modern eras, violence was explicit and police and/or the army were used to perpetuate it. The site of violence was the human body and by inflicting pain on the body, the human beings were compelled to detach themselves from their inner beings. Herein, human beings lose control over their consciousness. After surrendering one's being, one becomes a mindless object to be manipulated. In modern times, this brutality has been replaced by hegemony. Hegemony is exercised both upon institutions and humans. Hegemony is a structural phenomenon that refers to an organized assemblage of meanings and practices, values and actions [44]. Hegemony is not a 'passive mirror but an active force, one that also serves to give legitimacy to economic and social forms and ideologies so intimately connected to it' [44].

Consumerism is a global phenomenon. It has created a world that transcends national, ethnic, religious, gender or economic identity. It has also transcended individuality. It has created a trans-individual world. In this world, individuals cannot experience their subjectivity. When individual experience is denied, self-identity remains attached with the collective consciousness. In this way, on the one hand, consumerism 'liberates' humanity from perennial global conflicts but at the same time, it enslaves it by snatching humanity at an individual level. 'Just as modern mass production requires the standardization of commodities, so the social process requires standardization of man and this standardization is called equality' [45].

The material world in which we live is the achievement of collective human labour. This revolution happened as an individual became part of the collectivity. The aspirations and concrete efforts of an individual culminated into an impersonal material world. In this process of collective endeavours, an individual was collectivized as a species-being. A species being is integrated into the collectivity. This phenomenon has been elaborated by a Punjabi mystic poet Bulleh Shah. His verses can be paraphrased as follows:

> Oh' Bullah, I am not going to die. The body you are burying is not me'.

This means that with the death of a person humanity does not die; it lives on. The phenomenon of humanity prevails across time and space. The phenomenon of 'species being' manifests through individuals. In this relationship, Nature remains the crucial partner because species beings came into existence with the interaction with Nature.

In primitive times, human beings identified themselves with Nature as well as with fellow humanity. The more the human race separates itself from the natural relationship of species being, the more it loses its essence and becomes harmful to itself and to Nature. Freire's basic assumption is that 'man's ontological vocation ... is to be a subject who acts upon and transforms his world, and in so doing moves toward ever new possibilities of fuller and richer life individually and collectively. This world to which he relates is not a static and closed order, a given reality which man must accept and to which he must adjust, rather, it is a problem to be worked on and solved [46].

The capitalist mode of production has destroyed the consciousness of species being and hence its relationship with collectivity. It has atomized human beings and detached them from collective responsibilities. This suits the capitalist system. Individuals think that their existence has one dimension and that is to buy and create a spectacle of buying. This consciousness promotes the spree of consumerism and increases the production of objects. Under consumerism, an individual remains a 'species being' by imbibing the stimulus of buying from collectivity but it does not sustain the historical reciprocal relationship with Nature and fellow human beings.

Sustainability is not a moral issue. This issue is related to the very existence of human beings. By denying the consciousness of species beings, humankind is denying the essence of humanity. A consumer has a body of a human being but he/she lacks the qualifications of a human.

Consumers lack the dispositions of critical thinking (open-mindedness and fair-mindedness) and therefore follow the crowd slavishly. This situation is not problematic for them. Snatching humans' critical thinking is snatching their freedom to choose. This is as lethal as physical violence. In both ways, the victim is forced to surrender their self-autonomy and freedom to life choices. Under Fascism, violence is authorized by the regime but in the case of consumerism, it is authorized by the collectivity. When this violence is streamed in the information networks, it becomes routinized. So, victims are silently dehumanized and nobody shows any moral indignation or remorse on them. Violence becomes a way of life.

Such a situation requires critical consciousness because 'the awakening of critical consciousness leads the way to the expression of social discontents precisely because these discontents are real components of an oppressive situation' [46]. Living in harmony is an act of giving and accommodating others. People who themselves are dependent on others do not have the capacity to give. Giving is an attitude that comes to a person who is independent. People who indulge in exploiting others or have psychic emptiness and keep on hoarding lose their capacity to give. Similarly, people with narcissistic tendencies lose the capacity to live in harmony with others. 'This act of giving always implies certain basic elements and these are care, responsibility, respect and knowledge' [45].

Under the pedagogy of consumerism, life has become beyond human control. 'What was once considered humanly possible, a question involving values and human ends was now reduced to the issue of what was technically possible' [47]. This has promoted the values and culture of positivism. In this culture, values are separated from knowledge and methodological inquiry.

Giroux explains this culture as 'not only are 'facts' looked upon as objective, but the researcher himself is seen as engaging in value-free inquiry, far removed from the untidy world of beliefs and values' [47]. This culture permeates every sphere of life and people define every interaction with life and people from this lens. The major problem with this culture is that it 'freezes both human beings and history' [47].

8.2.4 Social Unsustainability

Right from the beginning, the ruling class which is always a minority has been afraid of the majority. Earlier, the majority was terrorised by explicit punishments and in modern times it has become invisible and being indoctrinated from within. The majority does what the minority wants it to do. This dictation is conveyed so 'nicely' that the majority acts upon it sincerely and takes pride in it. Living under authority has been engraved in the unconscious of mankind. By doing this, an individual gives up 'the independence of one's own individual self and to fuse one's self with somebody or something outside oneself in order to acquire the strength which the individual self is lacking' [48]. In authoritarian societies, ideas pertaining to human's innate frailty or disposition to sin and deficiencies are propagated to develop a sense of inferiority in people. Authority makes life easy for the people by prescribing ways for every sphere of life. Religions are the best example of prescription. They offer guidance from birth to death in the minutest details or rituals. The believers don't have to even think about the veracity of these canons. The innate urge to live under authoritarian systems is depicted in a story written by an Egyptian writer, Yusuf Idris [49]. This story has been summarised as follows:

> **'The Chair Carrier**: *I saw a frail man carrying a huge imperial chair on his shoulders. He was passing through Republic Street in Cairo. Due to perspiring for centuries, his veins were turned into drains. He is looking for the Authority who may allow him to put this chair on the ground. He claims that he is looking for this permission ever since the river Nile was not named yet. I looked at the misery of this man and yelled at him to drop this chair at once. Chairs are made so that these carry people not that people carry them on their shoulders. He asks me to show him the authority to do so. I, incidentally, found an inscription on the front side of this chair. It reads that the chair-bearer has been carrying this chair for so long and the time has come that now the chair should carry him*

instead. I read this message to him. He was gazing at me in disbelief and said, 'I cannot read. I said but, I have read it for you.' He said, 'I could have believed you if you had shown me the authority to do so.' He left perspiring and making whimpering sounds with a load of an archetypal chair on his shoulders'.

8.3 Higher Education for Sustainable Development Goals (SDGs)

SDGs are related to the three dimensions of sustainability i.e. economy, society and environment. It is important to note that the majority of the SDGs have their roots in the problems related to capitalism and consumerism. For example, SDG1 (eliminating poverty); SDG 2 (zero hunger), SDG3 (ensure healthy lives and promote well-being for all at all ages); SDG4 (quality education); SDG 6 (clean water and sanitation) and SDG 10 (reduced inequalities) are connected through the element of income. Wealthy people and nations can get healthy food, have access to good health, sanitation and educational facilities. On the other hand people, living in poverty are denied all these things. The major cause of poverty is a global capitalist system. Robinson [50] argues that the growth of wealth and of poverty are two sides of the same coin. He further explains the relationship between poverty and the global capitalist system as:

> In the logic of global capitalism, the cheapening of labor and its social disenfranchisement by the neo-liberal state became conditions for 'development'. The very drive by local elites to create conditions to attract transnational capital has been what thrusts majorities into poverty and inequality (p. 361).

Higher education is an important level of education with reference to eliminating poverty and inequalities. It contributes toward better economic prospectus for the individuals and the society. Literature also indicates a correlation between people's earnings and university degrees [51, 52, 53]. Higher education institutions may provide equitable opportunities to economically deprived students to gain higher education and ultimately increase their family income. Another important role of universities in this regard would be to produce solution-based research to address the issue of poverty. There is a need of including transdisciplinary content in university courses to allow students to understand the reasons for economic inequalities within and across the countries by looking at the bigger and complex picture of global capitalism.

SDG 13 (climate action); SDG 14 (life below water); SDG 15 (life on land) are related to the environmental dimension of sustainability. These goals cannot be achieved without achieving SDG 12 i.e. responsible consumption and production because, in the presence of unsustainable consumption and production, life on land and below water will be at constant risk. One of the targets of the goal of 'responsible consumption and production' is to transform people's lifestyles. The target 12.8 is: 'By 2030, ensure that people everywhere have the relevant information and awareness for sustainable development and lifestyles in harmony with nature' [54]. Universities can play a role in sustainable production by partnering with industries to produce innovative and sustainable technologies and products. Moreover, universities need to raise awareness for sustainable consumption and lifestyles because higher education is a key driver to transform people's lifestyles to make them more sustainable [55].

The goals related to the social aspects of sustainability such as SDG 5 (gender equality), SDG 8 (decent work and economic growth) and SDG 16 (peace, justice and strong institutions) also require a pro-active role of higher education institutions. University teachers are the 'public intellectuals' [7] and universities are the sites for knowledge production. Universities can contribute towards SDG 5, SDG 8 and SDG 16 by actively holding dialogue and research on gender-related issues and just work. Critical dialogue on societal issues (at macro and micro levels) is a key element of critical pedagogy [46], social justice pedagogy [56] and pedagogy for sustainability education [25]. It is also perceived as a key element of pedagogy of harmony with Nature too.

8.3.1 Pedagogy of Harmony with Nature (PHN)

To understand the concept of PHN, it is important to define Nature. Williams [57] note that Nature is mostly assigned three meanings i.e. Nature as the inherent quality of something; Nature as the universal force and Nature as the external, material world. Are humans inside or outside Nature? Is an important question in this regard. Demeritt [58] believes that the status of humans with reference to Nature is historical and evolutionary. Primitive people are often viewed as a part of Nature since they were subject to the universal laws of Nature. However, modern humans are imagined as having escaped the biological imperatives of Nature and are viewed as external to Nature.

Sustainability is fundamentally a historical issue and necessitates a pedagogy that may promote the pedagogy built around historicity. Unlike PHN, the pedagogy of consumerism promotes an ahistorical relation with Nature. Giroux [47] explains the historicity of the pedagogy of consumerism by insisting that history has been stripped of its critical and transcendent content and can no longer provide society with the historical insights necessary for the development of collective critical consciousness'. As a result of the ahistorical nature, consumerism does not abide by any values. The relationships of an individual with nature, the physical world and fellow humans are not spontaneous and creative, these are prescriptive and followed without any critical reflections upon them. History or genealogy builds memory that helps in giving a perspective to life, events, people and even to ordinary objects. Freire claims that history provides possibilities. He maintains that 'history is a time filled with possibility and not inexorably determined — that the future is problematic and not already decided, fatalistically' [46].

History depicts humankind's relations with Nature and the material world created by these relations. This is the essence of human consciousness. This method of understanding human consciousness is called dialectic. Dialectical thought reveals the power of human activity and human knowledge as both a product and force in the shaping of social reality. Furthermore, 'dialectical thought argues that there is a link between knowledge, power, and domination. Thus it is acknowledged that some knowledge is false and that the ultimate purpose of critique should be critical thinking in the interest of social change' [47].

Critical dialogue is a key process of sustainability teaching [59]. It is a means of constructing collective knowledge related to different dimensions of sustainable development (economy, society and environment) and their complex interactions. Dialogue is particularly important to concrete new possibilities. Friere [46] explains it as: 'dialogue is never an end in itself but a means to develop a better comprehension about the object of knowledge' (p. 18). Dialogue is a kind of search for harmony. Like friendly, harmonious musical notes that merge together to create a new composition, in the process of creating a dialogue, individuals come together, listen to each other and construct knowledge. In this process, they accommodate each other by recognising and respecting their experiences and ideas. In a dialogue, individuals leave their egocentric positions and come forward into a third space. If a dialogue comes to a conclusion, individuals create a third opinion which surely includes individuals' views point but take a different form as in music, different notes merge together and create a new composition.

A prerequisite for critical dialogue is to keep it within the historical cause and effect relationship. For example, in traditional societies, a large percentage of women experience physical and psychological violence [60]. This practice has been going on for centuries but family break-up is almost negligible. The custodians of these cultures think this is the success of their culture. But the cost of this success is given by the obedient women. 'Obedience to a person, institution or power (heteronomous obedience) is submission; it implies the abdication of my anatomy and the acceptance of a foreign will or judgment in place of my own' [61]. Disobedience is the only effective tool that can dismantle the defense mechanism of any dominating culture or system. When people disobey they loudly proclaim that they are not afraid of freedom. They declare that they do not need any prescription for life. Such tendencies would break the status quo and create conflicts. 'The ... conflicts are the systematic products of the changing structure of a society and by their very nature tend to lead to progress. The 'order' of society, hence, becomes the regularity of change. The 'reality' of society is conflict and flux, not a 'closed' functional system' [44].

Educational institutions mostly produce the type of knowledge (as a kind of commodity) that is needed to maintain the dominant economic, political and cultural arrangement that exists [44]. The captives under this system have no role to play in creating this knowledge. They simply follow the set ideas without questioning them. Higher education can play a role as a key enabler to achieve SDGs by focusing on transforming the problematic meaning perspective regarding consumerism or the ideology of consumerism. Building on the ideas of Sterling [24], Vare and Scott [20], Shephard et al., [26] and Kalsoom and Shah [25] regarding the purposes of sustainability education, the pedagogy of harmony with Nature (PHN) refers to the learning processes (critical reflection, dialogue, collaboration) that help in promoting critical thinking dispositions (open-mindedness, fair-mindedness) or transforming worldviews. The content of PHN will be transdisciplinary focusing on the society, economy, environment and their complex interactions.

8.4 Conclusion

The PHN is hope for social change by transforming people's meaning perspectives ultimately leading to socio-economic and ecological justice and freedom within and across different communities. An individuals would be provided with a 'basis for the creative expression of his / her personality — in other words, to nurture free man who will be immune to manipulation and the exploitation of their suggestibility for the pleasure and profit of others' [48]. While becoming a consumer, individuals lose their own subjectivity. They become one-dimensional beings. PHN is expected to help people explore possibilities to regain their subjectivity or totality. This totality may create possibilities for new relationships with Nature as well as with fellow beings and social systems.

References

[1] Paula Gunn Allen. Iyani: it goes this way. In Geary Hobson, editor, *The Remembered Earth: An Anthology of Contemporary Native American Literature*, pp. 191–93. University of New Mexico Press, Albuquerque, 1979.

[2] Manuel Castells. Universities as dynamic systems of contradictory functions. In John Muller and Nico Cloete, *Challenges of globalisation. South African debates with Manuel Castells*, pp. 206–223. Maskew Miller Longman, Cape Town, 2001.

[3] Tristan McCowan. Opening spaces for citizenship in higher education: three initiatives in English universities. *Studies in Higher Education*, vol. 37, no. 1, pp. 51–67, 2012.

[4] Ira Harkavy. The role of universities in advancing citizenship and social justice in the 21st century. *Education, Citizenship and Social Justice*, vol. 1, no. 1, pp. 5–37, 2006.

[5] Anthony D. Cortese. The critical role of higher education in creating a sustainable future. *Planning for Higher Education*, vol. 31, no. 3, pp. 15–22, 2003.

[6] Kerry Shephard. Higher education for sustainability: seeking affective learning outcomes. *International Journal of Sustainability in Higher Education*, vol. 9, no. 1, pp. 87–98, 2008.

[7] Kerry Shephard. *Higher Education for Sustainable Development*. Palgrave Macmillan, London, 2015.

[8] Savo Heleta and Tohiera Moodien. SDGs and higher education–Leaving many behind. *University World News*, vol. 457, 2017. Retrieved from: www.universityworldnews.com/post.php?story=20170427064053237

[9] Marina Angelaki. An introduction to responsible research & innovation. *PASTEUR4OA*, 2016. Retrieved from: http://pasteur4oa.eu/sites/pasteur4oa/files/resource/RRI_POLICY%20BRIEF.pdf

[10] Purcell, Wendy Maria, Heather Henriksen, and John D. Spengler. Universities as the engine of transformational sustainability toward

delivering the sustainable development goals: "Living labs" for sustainability. *International Journal of Sustainability in Higher Education*, vol. 20, no. 8, pp. 1343–1357, 2019.

[11] Wu, Yen-Chun Jim, Shihping Huang, Lopin Kuo, and Wen-Hsiung Wu. Management education for sustainability: A web-based content analysis. *Academy of Management Learning & Education*, vol. 9, no. 3, pp. 520–531, 2010.

[12] Gerd Michelsen. Policy, politics and polity in higher education for sustainable development. In Matthias Barth, Gerd Michelsen, Marco Rieckmann, Ian Thomas, editors, *Routledge Handbook of Higher Education for Sustainable Development,* pp. 64–79. Routledge, New York, 2015.

[13] UN Department of Economic and Social Affairs. *Johannesburg Declaration on Sustainable Development*, 2002. Retrieved from: www.un.org /esa/sustdev/documents/WSSD_POI_PD/English/POI_PD.htm

[14] UNECE. UNECE Strategy for Education for Sustainable Development. Adopted at the High level meeting, 2005. Retrieved from: https://unece.org/esd-strategy

[15] Budd Hall and Rajesh Tandon. Community based participatory research and sustainable development goals. *Canadian Commission for UNESCO's IdeaLab*, 2017

[16] UNESCO. *Framework for the Implementation of Education for Sustainable Development (ESD)* beyond 2019. UNESCO, Paris, 2019.

[17] J. Felix Lozano, Alejandra Boni, Jordi Peris and Andrés Hueso. Competencies in Higher Education: A Critical Analysis from the Capabilities Approach. *Journal of Philosophy of Education*, vol. 46, no. 1, pp. 132–147, 2012.

[18] Arnim Wiek, Lauren Withycombe and Charles L. Redman. Key competencies in sustainability: a reference framework for academic program development. *Sustainability Science*, vol. 6, no. 2, pp. 203–218, 2011.

[19] Arran Stibbe. *The Handbook of Sustainability Literacy: Skills for a Changing World.* Independent Publishers Group, Chicago, 2009.

[20] Paul Vare and William Scott. Learning for a change: Exploring the relationship between education and sustainable development. *Journal of Education for Sustainable Development*, vol. 1, no. 2, pp. 191–198, 2007.

[21] Robert B. Stevenson. Schooling and environmental education: Contradictions in purpose and practice. *Environmental Education Research*, vol. 13, no. 2, pp. 139–153, 2007.

[22] Bob Jickling and Arjen E.J. Wals. Globalization and environmental education: Looking beyond sustainable development. *Journal of Curriculum Studies*, vol. 40, no. 1, pp. 1–21, 2008.
[23] Victoria Gill. Coronavirus: This is not the last pandemic. BBC, 2020. Retrieved from: www.bbc.com/news/science-environment-52775386
[24] Stephen Sterling. Transformative learning and sustainability: Sketching the conceptual ground. *Learning and Teaching in Higher Education*, vol. 5, no. 11, pp. 17–33, 2011.
[25] Qudsia Kalsoom and Ayesha Jalil Shah. Re-thinking sustainability teaching. In Walter Leal Filho, editor, *COVID-19: Paving the Way for a More Sustainable World, pages* pp. 427–441. Springer, Cham, 2021.
[26] Kerry Shephard, Qudsia Kalsoom, Ritika Gupta, Lorenz Probst, Paul Gannon, V. Santhakumar, Ifeanyi Glory Ndukwe and Tim Jowett. Exploring the relationship between dispositions to think critically and sustainability concern in HESD. *International Journal of Sustainability in Higher Education*, 2021.
[27] Jack Mezirow. How Critical Transformative Learning Reflection Triggers. In Jack Mezirow and Associates, *Fostering Critical Reflection in Adulthood*, pp. 1–20. Jossey-Bass Publishers, San Francisco, 1991.
[28] Edward W. Taylor. *The Theory and Practice of Transformative Learning: A Critical Review.* ERIC, Centre on Education and Training for Employment, Columbus, Ohio, 1998.
[29] Richard Kiely. A Chameleon with a Complex: Searching for Transformation in International Service-Learning. *Michigan Journal of Community Service Learning,* pp. 5–20, Spring, 2004.
[30] Robert Kegan. What "form" transforms? A constructive-developmental approach to transformative learning. In Knud Illeris, editor, *Contemporary Theories of Learning: Learning Theorists ...in their own words*, pp. 35–52. Routledge, New York, 2009.
[31] Antonio Negri and M. Hardt. *Empire*. Harvard University Press, Cambridge, 2000.
[32] Paul J. Crutzen. Nature. *Geology of mankind*, vol. 415, pp. 23, 2002.
[33] Will Steffen, Paul J. Crutzen and John R. McNeill. The Anthropocene: Are humans now overwhelming the great forces of nature? In Ross E. Dunn, Laura J. Mitchell and Kerry Ward, editors, *The New World History,* pp. 440–459. University of California Press, Berkeley, 2016.
[34] Eric Fromm. *Marx's Concept of Man.* Continuum, London, 1994.
[35] Eric Fromm. *To Have or To be?* Continuum, London, 1978.

[36] Henri Lefebvre and Donald Nicholson-Smith. *The Production of Space,* Blackwell, Oxford, 1991.
[37] Falk Pasi. *The Consuming Body.* Sage publications, London, 1994.
[38] Manuel Castells. *The Information Age.* John Wiley & Sons, New Jersey, 2010.
[39] Beiser, V. (2019).Why the world is running out of sand? BBC, 2019. Retrieved from: www.bbc.com/future/article/20191108-why-the-world-is-running-out-of-sand
[40] World Resources Institute. (2020). Global Emissions of CO2 From Fossil Fuels: 1900–2004. Washington, 2020. Retrieved from: www.wri.org/resources/charts-graphs/global-emissions-co2-fossil-fuels-1900-2004
[41] Parker (2018). Planet or Plastic? National Geographic, 2018. Retrieved from: www.nationalgeographic.com/magazine/2018/06/plastic-planet-waste-pollution-trash-crisis/
[42] Parker (2020). The world's plastic pollution crisis explained. National Geographic, 2020. Retrieved from: www.nationalgeographic.com/environment/habitats/plastic-pollution/
[43] Harriet Constable. Your brand new returns end up in landfill. BBC, 2019. Retrieved from: www.bbcearth.com/news/your-brand-new-returns-end-up-in-landfill
[44] Michael Apple. *Ideology and Curriculum.* Routledge, London, 2002.
[45] Eric Fromm. *The Art of Loving.* Bentham Books, New York, 1972.
[46] Paulo Freire. *Pedagogy of the Oppressed.* Continuum, London, 2005.
[47] Henry Giroux. *Pedagogy and the Politics of Hope.* Westview Press, Oxford, 1997.
[48] Eric Fromm. *The Fear of Freedom.* ARK paperbacks, London, 1984.
[49] Yousaf Idris. The chair carrier. In Denys Johnson-Davies, editor, *Homecoming, Sixty Years of Egyptian Short Stories,* pages 194–1998. The American University Cairo Press, Cairo, 2012.
[50] William I. Robinson. Global capitalism theory and the emergence of transnational elites. *Critical Sociology*, vol. 38, no. 3, pp. 349–363, 2012.
[51] Fischer, Claude S., and Michael Hout. Century of difference: How America changed in the last one hundred years. *Russell Sage Foundation*, New York, 2006.
[52] Hout, M. (2012). Social and economic returns to college education in the United States. *Annual Review of Sociology*, vol. 38, pp. 379–400, 2012.

[53] Kingston, Paul W., Ryan Hubbard, Brent Lapp, Paul Schroeder, and Julia Wilson. Why education matters. *Sociology of Education,* vol. 76, no. 1, pp. 53–70, 2003.

[54] United Nations. Goal 12: Ensure sustainable consumption and production patterns. Retrieved from: www.un.org/sustainabledevelopment/sustainable-consumption-production/

[55] César Tapia-Fonllem, Blanca Fraijo-Sing, Víctor Corral-Verdugo and Anais Ortiz Valdez. Education for sustainable development in higher education institutions: Its influence on the pro-sustainability orientation of Mexican students. *Sage Open,* vol. 7, pp. 1, 2017.

[56] Qudsia Kalsoom, Naima Qureshi, Maria Shiraz and Mashail Imran. Undergraduate research: A vehicle of transforming epistemological beliefs of preservice teachers. *International Journal of Interdisciplinary Educational Studies.* vol. 16, no. 2, 2021.

[57] Raymond Williams. *Keywords: A Vocabulary of Culture and Society*. Flamingo, London, 1983.

[58] David Demeritt. What is the 'social construction of nature'? A typology and sympathetic critique. *Progress in Human Geography*, vol. 26, no. 6, pp. 767–790, 2002.

[59] Qudsia Kalsoom and Naima Qureshi. Impact of sustainability-focused learning intervention on teachers' agency to teach for sustainable development. *International Journal of Sustainable Development & World Ecology*, pp. 1–13, 2021. https://doi.org/10.1080/13504509.2021.1880983

[60] Devries, Karen M., Joelle YT Mak, Claudia Garcia-Moreno, Max Petzold, James C. Child, Gail Falder, Stephen Lim, L.J. Bacchus, R.E. Engell, L. Rosenfeld, C. Pallitto, T. Vos, N. Abrahams and C. H. Watts. The global prevalence of intimate partner violence against women. *Science*, vol. 340, no. (6140), pp. 1527–1528, 2013.

[61] Eric Fromm. *On Disobedience and Other Essays*. Routledge, London, 1984.

Index

21st century 49

A

AASHE 25, 55
Academic life 2
Academic revolutions 26
Academic technical training 97
Access equity 91
Accreditation models 98
Action from universities 147
Advanced knowledge 108
Adventurers 18
Affordable and Clean Energy 31, 57
AHELO 97
Analytical qualities 173
Association for the advancement of sustainability in Higher Education 25

B

Beings 174, 181
Bio-environmental engineering 151
Biopolitical goals 186,
Brazilian HE systems 86
Brazilian higher education 83
Brundtland Report 84, 144

C

Capitalism 185
CCS 2019 53, 73
Challenges 23, 44
Changes 93, 108
Charting 15
Climate action 57, 146
Climate change 63, 71, 143, 189
Climbing adventure 5, 12
Cognitive reflection 119
Collaborative networks 54
Commitment 24, 57, 86, 158
Communication industries 186
Communities 31, 57, 117
Competence 16
Competencies 100, 154, 173
Competitiveness 109, 118
Comprehensive guidelines 50
Conceptual knowledge 6, 13
Conceptual terrain 12, 18
Confidence 108
Connections 5, 26
Course curriculum 85
COVID crisis 28
COVID-19 health crisis 74
Critical thinking dispositions 182, 196
Cultural achievement 84
Curiosity 131
Current generation 84

D

DESD Milestones 46
Detrimental frames 184

Development 23
Development of skills 122, 134
Development paradigm 44, 73
Different industries 149
Disjuncture 6
Dissemination
of knowledge 45, 54
Distance education 87
Distance learning 84, 89, 99
Distance learning higher
education 89
Distribution of
employements 149
DLHE 89, 101

E

Earth and environmental
engineering 151
Earth system science 151
EAUC 25
ECIU University 133
Eco-friendly products 146
Ecological engineering 151
Economic aspects 73, 145
Economic development 27, 84
Economic dimension 53, 67
Economy 36, 175, 193
Educating students 171
Education 37, 45, 57
Education concept 99
Education for All 84, 108
Education for sustainability 1
Education for Sustainable
Development 108, 141, 144, 155
Education policy 84
Education system 109
Educational approach 4, 18
Educational objectives 95
EESD 156
EFA 84
EfS 15
EFS knowledge mountains 15
Emotional education 172
Emotional intelligence 108, 171
Emotional intelligence abilities 173
Emotions 171, 177
Empathy 174
Empower people 84
Energy 57, 71, 125
Energy for Sustainability 62
Energy intensity 119
Energy-environmental aspect 122
Engagement 6, 34, 49
Engineering education for
sustainable development 156
Engineers 142, 157
ENQA 98
Entrepreneurial minded student 151
Entrepreneurial thinking 151
Entrepreneurship 24, 36, 85
Environment 31, 62
Environmental aspects 153
Environmental association for
universities and colleges 25, 64
Environmental degradation 144, 152
Environmental dimension 53, 73
Environmental engineering 120, 142,
150
Environmental engineers 147, 149
Environmental issues 148, 157, 174
Environmental protection 157
Environmental Protection
Agency 147
Environmental regulations 148
Environmental resource
engineering 151
Environmental sanitary
engineering 151

Environmental sustainability 174
Environmental systems engineering 151
EPA 147
EQAR 98
ERASMUS+ 128, 138
ESD 147, 154, 183
Ethics for sustainability 54
Ethnic quota law 87
Ethnicity 91, 104
Experts' knowledge 6
External leadership 47

F

Faculty of Engineering of the University of Porto 45, 64
Fair-mindedness 191
Fast-changing professional environment 147
FEUP 45, 62
Fies 87, 94
Financial issues 175
Future 47, 70, 87
Future generation 142, 174

G

Gen Z 146
Gender 91, 190
GHG 119, 143
Global challenges 153
Global communities 153, 183
Global Compact 24, 32
Global compact principles 24, 31
Global Reporting Initiative 25, 143
Governance 36, 47, 64
Governments 45, 74, 85, 147
Graduated-level employees 95
Green campus activities 34
Greenhouse gas emissions 118, 150
GreenMetric World Universities ranking 45, 62, 73
GRI 25, 50, 66
GRI disclosures 25
GRI guidelines 25, 50, 66
Guiding principles for engineers 157

H

HE quality 98
HE Student Financing Fund 87
Healthy environment 34, 148
HEI 50
HEIs 56, 64, 88, 104
HESD 182
Hidden complexity 4, 18
Hierarchical learning 8
Higher Education 27, 49, 56
Higher education courses 119
Higher Education Institutions 24, 118, 193
History of SD 141
Human behaviour 185
Human rights 24, 85, 104, 144
Humanity 32, 153, 190

I

IIRC 50
Inclusive societies 44
Incorporate students 134
Incorporating mobility 120
Industries 146, 194
Industry sustainability practices 149
Inequalities in class 91
Inequality 152, 193
Initiatives of universities 35
Innovation 17, 48, 67, 74
Innovation activity 24

Innovative 118, 147, 194
Institutional commitment 54
Integrative assessment 99
Intercultural dimensions 136
Inter-cultural solutions 154
International sustainable campus network 25
Internationalization at Home 136
Involvement 23, 48, 71, 134
ISCN 25
ISCTE-UIL 45, 62

J
Johannesburg World Summit on SD 33

K
Knowledge 84, 94, 108, 148
Knowledge based-economy 29, 156
Knowledge integrations 19
Knowledge mountains 9, 15
Knowledge structures 9, 19

L
Labor market 86, 100
Learn from failure 151
Learners 134, 156
Learning 2, 47, 184
Learning activities 48
Learning process for adults 120
Life on Land 58, 194
Lifelong learning 57, 106
Living in harmony 191
Local pollutants 119
Loss of biodiversity 144, 152

M
Mapping 4, 7, 31
Mass education 86, 95, 104
MDGs 28, 44, 73
Meaning-making 6
Micromodules 133
Millennial 146
Millennium Declaration 84
Millennium Development Goals 28, 84, 143
Millennium Summit 84
Mission of universities 26, 35
Mobility 48, 117, 121, 138
Mobility on sustainable cities 118
Monitoring students 108
Motivation 49, 108, 174
Mountain guides 5, 18
Multicultural experiences 137
Multiple functions of universities 26

N
National development and capacities' 44
National Student Assistance Plan 87, 105
Natural resources 44, 71, 143, 171
Nature 13, 143, 181, 188, 194
Non-university institutions 85, 86

O
OECD 97, 114
On campus 47, 57, 89, 102
Open-mindedness 191, 196

P
Partnerships 27, 54, 58, 143, 183
Peace 44, 58, 104, 153, 194
Peace studies 85
Pedagogic resonance 1, 6, 8, 18
Pedagogical modules 123
Pedagogy 2, 18, 104, 181, 194

Pedagogy of domination 189
Pedagogy of harmony 185, 194, 196
Pedagogy of sustainability 182
People 11, 49, 104, 176, 182, 197
Perceptions 10, 156
PHN 194, 196, 197
Planet 17, 44, 109, 141, 185
Players 23, 37, 47
Plunder of nature 188
Pnaes 87, 105
Political agenda 84, 91
Portugal 43, 50, 90, 130
Portuguese 43, 49, 55, 56, 61, 62
Portuguese HEIs 43, 52, 55, 59, 61, 62
Portuguese students 137
Poverty 37, 57, 143, 193
Power of internationalization 136
Problem-solving skills 108
Procedural knowledge 6, 13
Professional life 108
Professional training in HE 94
Project-based learning 120, 124, 128
Prosperity 44, 72, 142, 156
Prouni 87, 94
Pull 141, 147, 159
Pull from industry 142, 147, 159
Push 141, 146, 159
Push from students 146, 147, 159

Q

Quality assessment 84, 91, 93, 97, 106
Quality education 27, 31, 45, 57, 103, 193
Quality of life 35, 104, 143, 156
Quick thinking 147

R

Race 23, 55, 144, 190
Race for sustainable development 23
Recycling bin 188
Research 18, 57, 107, 118, 120, 134, 173
Research activities 48, 120, 134
Resilient communities 134
Responsible Consumption and Production 31, 57, 182
Responsible management education 31, 32
Role 3, 13, 24, 141, 148, 182
Role of engineers 156

S

Sanitary engineering 151
Scientific leaders 94
SD 44, 45, 47
SD principles 154, 157, 158
SDG 28, 57, 104, 159, 193
SDG 11 57, 70, 117,
SDG 16 59, 194
SDG 17 45, 58
SDG 2 57, 193
SDG 2030 Agenda 28
SDG 3 57, 58, 69
SDG 4 37, 84, 104
SDG 8 45, 194
SDG 9 57, 69, 70
SDG impact 50
SDG1 182, 193
SDG 10 45, 57, 193
SDG 12 57, 182, 194
SDG 13 57, 182
SDG 14 57, 69, 194

SDG 15 58, 194
SDG 5 57, 194
SDG 6 57, 196
SDGs 29, 32, 33, 142, 183
Segmented learning 21
Self-awareness 174
Self-development 95
Self-management 174
Self-regulation 37, 98
Skilled labour force 182
Skills 15, 74, 108, 119, 154
Skillset 147, 151
Smart and sustainable mobility 118, 120, 134
Smart skills 154
Social aspects 102, 194
Social dimension 53, 73
Social inclusion 91, 104, 118
Social issues 145, 172, 174
Social skills 174
Social systems 197
Social transformations 91
Social unsustainability 192
Society 2, 44, 84, 173, 174
Sophisticated instructional discourse 1, 4, 18
Stakeholders 17, 119, 150
STARS 45, 56
Student 6, 85, 102, 104, 135, 171
Student enrollments 88
Student sustainability initiatives 146
Student-facilitated programs 147
Students' thinking 184
Students' worldviews 184
Sustainability 1, 43, 148, 157, 174, 181
Sustainability assessment 49
Sustainability capabilities 183
Sustainability competencies 157, 183
Sustainability consciousness 183
Sustainability courses 183
Sustainability education 141, 145, 154, 183
Sustainability engineer 148
Sustainability in higher education 25, 45, 181
Sustainability initiatives 146
Sustainability learning 183
Sustainability literacy 183
Sustainability rankings 43, 55
Sustainability report 25, 45, 62
Sustainability reporting 43, 69
Sustainability skills 156
Sustainability Tracking, Assessment and Rating System 55,
Sustainable Campus Conference 73
Sustainable Campus Network 25, 54
Sustainable Cities and Communities 31, 57, 117, 134
Sustainable companies 150
Sustainable development 17, 23, 31, 141, 172, 194
Sustainable development concept 84
Sustainable development goal 4 83
Sustainable Development Goals 25, 29, 85, 119, 171, 173, 193
Sustainable development solutions network 25, 47
Sustainable development through curricula 33, 35
Sustainable living 146, 153, 159
Sustainable mobility 117, 120, 134
Sustainable practices 35, 43, 146, 155
Sustainable urban environments 134
Sustenance of our planet 156
Syllabus 107, 119, 123

T
Teaching 2, 88, 103, 120, 122, 195
Teaching and research activities 120
Teaching methods 2, 3, 5, 13, 128
Teaching process 136
Teaching-learning 53, 65, 119, 136
Technical knowledge 148, 151
Technical-scientific report 133
Technological challenge 119
Technological innovations 85
THE 45, 49, 55, 57, 59–61, 183
Threshold concepts 1, 11, 23
Times Higher Education 56, 109, 183
Tools 15, 123, 189
Training for sustainability 54
Training in research 95
Transformation of universities 24
Transformative learning 182, 184
Transformative pedagogy 185
Transport 117, 118, 120
Transport and mobility 134, 138
Transportation 63, 118, 120, 121, 128
Transversal competences 134

U
UC 45, 62, 69
UMinho 43, 45, 66
UN 25, 43, 119, 143, 183
UN Agenda 2030 85
UN Decade for Education for SD 33
UN Sustainable Development Goals 76, 119
Understanding 2, 10, 121, 156, 175, 195
UNECE strategy 183, 198
UNESCO 45, 69, 144
United Nations 24, 43, 117, 182
United Nations Decade of Education for Sustainable Development 46
Universities 3, 47, 57, 72, 106, 181
University for All Program 87
University institutions 85, 86
University internationalisation 28
University of Coimbra 54, 58, 59, 63, 68, 70
University of Minho 43, 59, 63, 66
Unsustainable practices 15, 146, 159
UNWCED 142
Upholding sustainable development goals 171, 172

V
Values 2, 5, 51, 141, 152, 157, 174, 190

W
Waste impacts 25
Wicked problems 1, 16

Y
Youth sustainability activism 146

About the Editors

Carolina Machado received her PhD degree in Management Sciences (Organizational and Policies Management area/Human Resources Management) from the University of Minho in 1999, Master degree in Management (Strategic Human Resource Management) from the Technical University of Lisbon in 1994, and Degree in Business Administration from the University of Minho in 1989. Teaching in the Human Resources Management subjects since 1989 at the University of Minho, she is since 2004 Associated Professor, with experience and research interest areas in the field of Human Resource Management, International Human Resource Management, Human Resource Management in SMEs, Training and Development, Emotional Intelligence, Management Change, Knowledge Management, and Management/HRM in the Digital Age/Business Analytics. She is the Head of the Human Resources Management Work Group at the School of Economics and Management at University of Minho, Coordinator of Advanced Training Courses at the Interdisciplinary Centre of Social Sciences, Member of the Interdisciplinary Centre of Social Sciences (CICS.NOVA.UMinho), University of Minho, as well as Chief Editor of the *International Journal of Applied Management Sciences and Engineering (IJAMSE)*, Guest Editor of journals, books Editor, and book Series Editor, as well as reviewer in different international prestigious journals. In addition, she has also published both as editor/co-editor and as author/co-author several books, book chapters and articles in journals and conferences.

João Paulo Davim is Professor at the University of Aveiro, Portugal. He is also distinguished as honorary professor in several universities/colleges/institutes in China, India, and Spain. He received his Ph.D. degree in Mechanical Engineering in 1997, M.Sc. degree in Mechanical Engineering (materials and manufacturing processes) in 1991, Mechanical Engineering degree (5 years) in 1986, from the University of Porto (FEUP), the Aggregate title (Full Habilitation) from the University of Coimbra in 2005 and the D.Sc. (Higher Doctorate) from London Metropolitan University in

2013. He is a Senior Chartered Engineer by the Portuguese Institution of Engineers with an MBA and Specialist titles in Engineering and Industrial Management as well as in Metrology. He is also Eur Ing by FEANI-Brussels and Fellow (FIET) of IET-London. He has more than 30 years of teaching and research experience in Manufacturing, Materials, Mechanical, and Industrial Engineering, with special emphasis in Machining & Tribology. He has also interest in Management, Engineering Education, and Higher Education for Sustainability. He has guided large numbers of postdoc, Ph.D. and master's students as well as has coordinated and participated in several financed research projects. He has received several scientific awards and honors. He has worked as evaluator of projects for ERC-European Research Council and other international research agencies as well as examiner of Ph.D. thesis for many universities in different countries. He is the Editor in Chief of several international journals, Guest Editor of journals, books Editor, book Series Editor and Scientific Advisory for many international journals and conferences. Presently, he is an Editorial Board member of 30 international journals and acts as reviewer for more than 100 prestigious Web of Science journals. In addition, he has also published as editor (and co-editor) more than 200 books and as author (and co-author) more than 15 books, 100 book chapters and 500 articles in journals and conferences (more than 300 articles in journals indexed in Web of Science core collection/h-index 61+/12500+ citations, SCOPUS/h-index 66+/15500+ citations, Google Scholar/h-index 85+/25500+ citations). He has been listed in World's Top 2% Scientists by Stanford University study.

For Product Safety Concerns and Information please contact our EU
representative GPSR@taylorandfrancis.com
Taylor & Francis Verlag GmbH, Kaufingerstraße 24, 80331 München, Germany

www.ingramcontent.com/pod-product-compliance
Ingram Content Group UK Ltd.
Pitfield, Milton Keynes, MK11 3LW, UK
UKHW021821240425
457818UK00001B/17

* 9 7 8 8 7 7 0 2 2 4 3 1 4 *